T0251659

Gasification Technologies

A Primer for Engineers and Scientists

CHEMICAL INDUSTRIES

A Series of Reference Books and Textbooks

Consulting Editor

HEINZ HEINEMANN
Berkeley, California

Gasification Technologies

A Primer for Engineers and Scientists

John Rezaiyan
Nicholas P. Cheremisinoff

Taylor & Francis
Taylor & Francis Group

Boca Raton   London   New York   Singapore

A CRC title, part of the Taylor & Francis imprint, a member of the
Taylor & Francis Group, the academic division of T&F Informa plc.

The source for the cover picture is the U.S. Department of Energy, National Energy Technology Laboratory's
Web site: http://www.netl.doe.gov/cctc/resources/database/photos/phototampa.htm

Published in 2005 by
CRC Press
Taylor & Francis Group
6000 Broken Sound Parkway NW, Suite 300
Boca Raton, FL 33487-2742

Library of Congress Cataloging-in-Publication Data

Catalog record is available from the Library of Congress

Taylor & Francis Group
is the Academic Division of T&F Informa plc.

**Visit the Taylor & Francis Web site at
http://www.taylorandfrancis.com**

**and the CRC Press Web site at
http://www.crcpress.com**

Table of Contents

Preface

Gasification technologies offer the potential of clean and efficient energy. The technologies enable the production of synthetic gas from low or negative-value carbon-based feedstocks such as coal, petroleum coke, high sulfur fuel oil, materials that would otherwise be disposed as waste, and biomass. The gas can be used in place of natural gas to generate electricity, or as a basic raw material to produce chemicals and liquid fuels.

Gasification is a process that uses heat, pressure, and steam to convert materials directly into a gas composed primarily of carbon monoxide and hydrogen. Gasification technologies differ in many aspects but rely on four key engineering factors:

1. Gasification reactor atmosphere (level of oxygen or air content)
2. Reactor design
3. Internal and external heating
4. Operating temperature

The feedstock is prepared and fed, in either dry or slurried form, into a reactor chamber called a gasifier. The feedstock is subjected to heat, pressure, and either an oxygen-rich or oxygen-starved environment within the gasifier. All commercial gasifiers require an energy source to generate heat and begin processing.

There are three primary products from gasification:

- Hydrocarbon gases (also called syngas)
- Hydrocarbon liquids (oils)
- Char (carbon black and ash)

Syngas can be used as a fuel to generate electricity or steam, or as a basic building block for a multitude of chemicals. When mixed with air, syngas can be used in gasoline or diesel engines with few modifications to the engine.

Both pyrolysis and gasification convert carbonaceous materials into energy-rich fuels by heating the feedstock under controlled conditions. Whereas incineration fully converts the input material into energy and ash, these processes deliberately limit the conversion so that combustion does not take place directly. Instead, they convert the material into valuable intermediates that can be further processed for materials recycling or energy recovery.

Gasification in particular offers more scope for recovering products from waste than incineration. When waste is burned in a modern incinerator the only practical product is energy, whereas the gases, oils, and solid char from gasification can not only be used as a fuel but also be purified and used as a feedstock for petro-chemicals and other applications. Gasification can be used in conjunction with gas engines and gas turbines to obtain higher conversion efficiency than conventional fossil-fuel electric power generation. In contrast, conventional incineration, used in conjunction with steam-cycle boilers and turbine generators, achieves lower efficiency. Gasification can help meet renewable energy steam targets, address concerns about global warming, and contribute to achieving Kyoto Protocol commitments.

There are more than 150 companies around the world that are marketing systems based on gasification concepts. Many of these are optimized for specific wastes or particular scales of dedicated energy production operations. They vary widely in the extent to which they are proven in operation. In addition, there are more than 100 facilities operating around the world.

This book serves as a primer to coal and biomass gasification technologies. It is meant as an introduction and overview of current technology developments, and to provide readers with a general understanding of the technology challenges for large-scale commercialization. While there is an abundant source of literature both on the World Wide Web and in printed form, the information and experiences in development and commercialization are fragmented. This volume helps to place the technology and research and development challenges into perspective.

Nicholas P. Cheremisinoff, Ph.D.
A. John Rezaiyan
Princeton Energy Resources International, LLC

About The Authors

Nicholas P. Cheremisinoff has 30 years of industry and applied research and development experience throughout the petrochemical and allied industries. His assignments have focused on implementation of clean technologies for manufacturing and energy production, with experiences ranging from fossil energy to biomass and wind energy applications. He has worked extensively on overseas assignments for donor agencies such as the United States Agency for International Development, for international lending institutions including the World Bank Organization, and for numerous private sector clients. He is the author, co-author, or editor of more than 100 technical books. Dr. Cheremisinoff received his B.Sc., M.Sc., and Ph.D. degrees in chemical engineering from Clarkson College of Technology.

A. John Rezaiyan is Vice President for Advanced Engineering Group at Princeton Energy Resources International LLC (PERI). He has 25 years of experience in fluidized-bed combustion and gasification technology development. He works closely with technology developers, project developers, government agencies, and financial institutions to assess market potential and technical, economic, and commercial viability of advanced power generation, coke making, and still making technologies. More recently, Mr. Rezaiyan has focused his effort in helping clients to commercialize their technology. He is the author of a number of articles addressing the market potential of and financing strategies for advanced clean coal technologies. Mr. Rezaiyan received his B.S. degree in chemical engineering from University of Maryland at College Park.

1

Principles of Gasification

INTRODUCTION

Gasification technologies have been commercially applied for more than a century for the production of both fuels and chemicals. Current trends in the power generation and refinery industries support the observation that advanced stages of the technology will continue to be applied toward the synthesis of syngas, with an increasing number of applications in power generation, fuels, and basic chemicals manufacturing.

Attractive features of technology include:

- The ability to produce a consistent product that can be used for the generation of electricity or as primary building blocks for manufacturers of chemicals and transportation fuels.
- The ability to process a wide range of feedstocks including coal, heavy oils, petroleum coke, heavy refinery residuals, refinery wastes, hydrocarbon contaminated soils, biomass, and agricultural wastes.
- The ability to remove contaminants in the feedstock and to produce a clean syngas product.
- The ability to convert wastes or low-value products to higher value products.

- The ability to minimize the amount of solid waste requiring landfill disposal. Solid by-products have a market value can be used as fuel or construction material, and are non-hazardous.

This chapter provides an overview of the technology, the products that can be made using this technology, important terminology, and a general overview of the history and modern development trends.

HISTORICAL PERSPECTIVE AND COMMERCIALIZATION TRENDS

Historical Perspectives

The earliest practical production of synthetic gas (syngas) is reported to have taken place in 1792 when Murdoch, a Scottish engineer, pyrolyzed coal in an iron retort and then used the product, coal gas, to light his home.[1]

Later on, Murdoch built a gas plant for James Watt, the inventor of the steam engine, and applied the technology to lighting one of Watt's foundries.

The first gas company was established in 1812 in London to produce gas from coal and to light the Westminster Bridge. In 1816, the first gas plant for the manufacture of syngas from coal was built in the United States to light the streets of the city of Baltimore. By 1826, gas plants were also built to manufacture gas for lighting the streets of Boston and New York City. Soon thereafter, gas plants and distribution networks were built to light the streets of most major cities throughout the world.

In 1855, the invention of the Bunsen burner premixed air and gas, allowing it to burn more economically, at very high temperatures, and without smoke. This invention added impetus to the further use of gas.

---

[1]Lowry, H. H., editor, *Chemistry of Coal Utilization,* John Wiley & Sons, 1945, p. 1252.

In the latter half of the 19ᵗʰ century coal gasification became a commercial reality through the use of cyclic gas generators,[2] also know as air-blown gasifiers. By 1875, manufactured gas was being widely used for home lighting, and by the end of the century it was applied to domestic and industrial applications. In the United States more than 1200 gas plants were in operation by the late 1920s.

In early 1900s, biomass gasification processes were also widely used to manufacture synthetic gases for production of fuels, chemicals, and hydrogen. During World War II, over 1 million air-blown gasifiers were built to produce synthetic gas from wood and charcoal to power vehicles and to generate steam and electricity.[3]

After World War II, the discovery of large quantities of low-cost natural gas with heating values of about 37 MJ/m^3 (1000 Btu/ft^3) led to the demise of the synthetic gas manufacturing industry.

Renewed Interest and the Incentives for Commercialization

Interest in gasification technologies was renewed throughout the 1960s and 1970s when controversial projections suggested that natural gas reserves would be depleted and demand would exceed reserves by the 1980s and 1990s. Also the oil embargo of 1973 created awareness for the need to identify alternative sources of fuel.

Throughout the 1980s, researchers and industry came to recognize some of the environmental benefits of gasification technology. More restrictive and stringent environmental standards aimed at controlling power plant emissions, and domestic and industrial waste landfills, and an increased

---

[2]Cyclic gas generators converted coke, a by-product of high-temperature pyrolysis process, to a synthetic gas by alternatively exposing the coke to air to provide heat and to steam to produce a gas that burned with a blue flame. The coal gas was know as "blue water gas" (Probstein, R. F. and Hicks, R. E., *Synthetic Fuels,* McGraw-Hill, 1982, p. 7).

[3]Klass, D. L., *Biomass for Renewable Energy, Fuels, and Chemicals,* Academic Press, 1998, p. 271.

emphasis on greenhouse gas reductions provided incentives for both government and industry stakeholders to explore and promote the commercialization of gasification technologies.

Commercialization Growth and Today's Applications

Based on a survey reported for 2003, there are 163 commercial gasification projects worldwide consisting of a total of 468 gasifiers.[4] More than 120 plants began their operations between 1960 and 2000 with the majority (more than 72 plants) commissioned after 1980. Up to 34 new plants are at various stages of planning and construction.

The majority of the existing plants were designed and constructed to produce a synthetic gas, consisting primarily of hydrogen and carbon monoxide (CO), which is used for the production of hydrogen or Fischer-Tropsche (F-T) syncrude. Hydrogen is then used to produce a wide variety of chemicals and fertilizers. The Fisher-Tropsch syncrude is used to man-ufacture transportation fuels, lube oils, and specialty waxes.

Among the most recent plants are those designed to produce a synthetic gas suited for firing in gas turbines for the production of clean electric power. Major projects in the United Sates for clean power generation have included Global Energy's Wabash River Power Station in Indiana and Tampa Electric's Polk County Power Station in Florida. These plants began operation in 1995 and 1996, respectively. They both use coal as their feedstock and are based upon an Integrated Gasification Combined Cycle (IGCC) plant configuration. The U.S. Department of Energy (DOE) and the private sector project owners shared the cost for design, construction, and initial operation of these plants.

Recent commercial projects use refinery waste or prod-ucts that no longer have a positive market value, such as petroleum coke (petcoke) or heavy oils. Many of these projects are referred to as "trigeneration" plants because they produce hydrogen, power, and steam for use within the refinery and

---

[4]U.S. Department of Energy, http://www.netl.doe.gov/coalpower/gasification/models.

for export. Gasification projects in the U.S. using petcoke or heavy oils as feedstock include the Frontier Oil gasification project in El Dorado, TX; Motiva gasification project in Delaware City, DE; Farmland Industries gasification project in Coffeyville, KS; and Exxon-Mobile gasification project in Baytown, TX.

European refinery gasification projects include the API project in Falconara, Italy; Sarlux project in Sardinia, Italy; ISAB project in Sicily, Italy; and Shell project in Pernis, Netherlands. Several plants are also reported to gasify biomass to produce gaseous fuels or electric power.

Examples of operating plants using biomass as feedstocks include Rudersdorfer Zement project in Germany; Lahden Lämpövoima Oy project, Corenso United Oy Ltd project, and Oy W. Schauman ab Mills project in Finland; Netherlands Refinery Company BV project in Netherlands; BASF plc project in United Kingdom; ASSI project and Sydkraft AB in Sweden; and Portucel project in Portugal. These projects are generally smaller (6 to 54 MW equivalent in size) than projects using coal, petcoke, or heavy oil as feedstock.

GASIFICATION PRINCIPLES

Overview

Gasification is a process for converting carbonaceous materials to a combustible or synthetic gas (e.g., H_2, CO, CO_2, CH_4). In general, gasification involves the reaction of carbon with air, oxygen, steam, carbon dioxide, or a mixture of these gases at 1,300°F or higher to produce a gaseous product that can be used to provide electric power and heat or as a raw material for the synthesis of chemicals, liquid fuels, or other gaseous fuels such as hydrogen.

Once a carbonaceous solid or liquid material is converted to a gaseous state, undesirable substances such as sulfur compounds and ash may be removed from the gas. In contrast to combustion processes, which work with excess air, gasification processes operate at substoichiometric conditions with the oxygen supply controlled (generally 35 percent of the

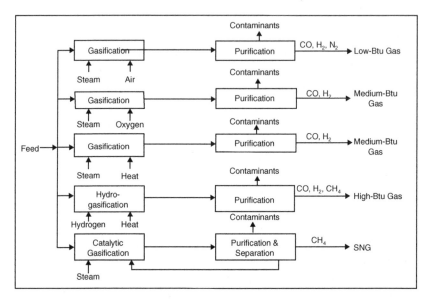

Figure 1.1 Gasification methods.

amount of O_2 theoretically required for complete combustion or less) such that both heat and a new gaseous fuel are produced as the feed material is consumed.

Some gasification processes also use indirect heating, avoiding combustion of the feed material in the gasification reactor and avoiding the dilution of the product gas with nitrogen and excess CO_2.

Figure 1.1 shows the principal methods for gasifying a carbonaceous material.

When a carbonaceous material is heated, either directly or indirectly, under gasification conditions, it is first pyrolyzed. During pyrolysis light volatile hydrocarbons, rich in hydrogen, are evolved and tars, phenols, and hydrocarbon gases are released. During pyrolysis the feedstock is thermally decomposed to yield solid carbon and a gas product stream that has higher hydrogen content than the original carbonaceous feed material.

Hydrogenation

In a gasification process the feedstock is hydrogenated. This means hydrogen is added to the system directly or indirectly or the feedstock is pyrolyzed to remove carbon to produce a product with a higher hydrogen-to-carbon ratio than the feedstock. These processes may be carried out separately or simultaneously. The more hydrogen that is added or the more carbon removed, the lower the overall efficiency of the synthetic gas production process. In an indirect hydrogenation process, steam is used as a hydrogen source and hydrogen is produced within the gasification reactor.

An indirect hydrogenation process that is still under development is catalytic gasification. In this process, a catalyst accelerates the gasification reactions, resulting in the formation of hydrogen and CO, at relatively low temperatures. This process also promotes catalytic formation of methane at the same low temperature within the same reactor. Catalyst deactivation and costs have been a major impediment to the commercialization of this process.

In a direct hydrogenation process feedstock is exposed to hydrogen at high pressures to produce a gas with higher methane content than indirect hydrogenation processes.

Indirect hydrogenation processes are also known as air or oxygen blown gasification, depending on whether air or oxygen is used as the oxidant source. If heat is also provided indirectly, air or oxygen is not used to combust some of the feedstock in the gasifier. This results in an increase in the reactor temperature to the desired gasification reaction temperatures, which is a process referred to as steam reforming. Direct hydrogenation processes are called hydro-gasification.

Stoichiometric Considerations

Depending on the gasification process, reactions that take place in a gasifier include:

$$(1)\ C + O_2 \rightarrow CO_2$$

$$(2)\ C + \tfrac{1}{2}O_2 \rightarrow CO$$

$$(3) \ H_2 + \tfrac{1}{2}O_2 \rightarrow H_2O$$

$$(4) \ C + H_2O \rightarrow CO + H_2$$

$$(5) \ C + 2H_2O \rightarrow CO_2 + 2H_2$$

$$(6) \ C + CO_2 \rightarrow 2CO$$

$$(7) \ C + 2H_2 \rightarrow CH_4$$

$$(8) \ CO + H_2O \rightarrow H_2 + CO_2$$

$$(9) \ CO + 3H_2 \rightarrow CH_4 + H_2O$$

$$(10) \ C + H_2O \rightarrow \tfrac{1}{2}CH_4 + \tfrac{1}{2}CO_2$$

Most of the oxygen injected into a gasifier, either as pure oxygen or air, is consumed in reactions (1) through (3) to provide the heat necessary to dry the solid fuel, break up chemical bonds, and raise the reactor temperature to drive gasification reactions (4) through (9).

Reactions (4) and (5), which are known as water-gas reactions, are the principal gasification reactions, are endothermic, and favor high temperatures and low pressures.

Reaction (6), the *Boudourd reaction,* is endothermic and is much slower than the combustion reaction (1) at the same temperature in the absence of a catalyst.

Reaction (7), hydro-gasification, is very slow except at high pressures.

Reaction (8), the *water-gas shift reaction*, can be important if H_2 production is desired. Optimum yield is obtained at low temperatures (up to 500°F) in the presence of a catalyst and pressure has no effect on increasing hydrogen yield.

Reaction (9), the *methanation reaction,* proceeds very slowly at low temperatures in the absence of catalysts.

Reaction (10) is relatively thermal neutral, suggesting that gasification could proceed with little heat input but methane formation is slow relative to reactions (4) and (5) unless catalyzed.

In addition to the gasification agent (air, oxygen, or steam) and the gasifier operating temperature and pressure, other factors affect the chemical composition, heating value, and the end use applications of the gasifier product gas. The following factors affect the quality of the product gas:

- Feedstock composition
- Feedstock preparation and particle size
- Reactor heating rate
- Residence time
- Plant configuration such as:
  - Feed system - dry or slurry
  - Feedstock-reactant flow geometry
  - Mineral removal system - dry ash or slag
  - Heat generation and transfer method — direct or indirect
  - Syngas cleanup system - low or high temperature and processes used to remove sulfur, nitrogen, particulates, and other compounds that may impact the suitability of the syngas for specific applications (i.e., turbine and fuel cell for electric power generation, hydrogen production, liquid fuel production, or chemical production).

Depending on the gasifier system configuration, operating conditions, and gasification agent, four types of synthetic gas can be produced:

- Low heating-value gas (3.5 to 10 MJ/m^3 or 100 to 270 Btu/ft^3) can be used as gas turbine fuel in an IGCC system, as boiler fuel for steam production, and as fuel for smelting and iron ore reduction applications. However, because of its high nitrogen content and low heating value, it is not well suited as a natural gas replacement or for chemical synthesis. Use of low heating-value gas for fuel cell applications also increases gas upgrading and processing costs, including compression costs if high pressure fuel cells are used.
- Medium heating-value gas (10 to 20 MJ/m^3 or 270 to 540 Btu/ft^3) can be used as fuel gas for gas turbines in IGCC applications, for substitute natural gas (SNG) in combination with methanation process, for hydrogen production, for fuel cell feed, and for chemical and fuel synthesis.

- High heating-value gas (20 to 35 MJ/m^3 or 540 to 940 Btu/ft^3) can also be used as fuel gas for gas turbines in IGCC applications, for SNG and hydrogen production, for fuel cell feed, and for chemical and fuel synthesis. However, it does not require as much upgrading and methanation to produce SNG.
- SNG (over 35 MJ/m^3 or 940 Btu/ft^3) can be easily substituted for natural gas and therefore is suitable for hydrogen and chemical production as well as fuel cell feed.

GASIFICATION VERSUS COMBUSTION

Comparisons of General Features

Gasification is not an incineration or combustion process. Rather, it is a conversion process that produces more valuable and useful products from carbonaceous material. Table 1.1 compares the general features of gasification and combustion technologies.

Both gasification and combustion processes convert carbonaceous material to gases. Gasification processes operate in the absence of oxygen or with a limited amount of oxygen, while combustion processes operate with excess oxygen.

The objectives of combustion are to thermally destruct the feed material and to generate heat. In contrast, the objective of gasification is to convert the feed material into more valuable, environmentally friendly intermediate products that can be used for a variety of purposes including chemical, fuel, and energy production. Elements generally found in a carbonaceous material such as C, H, N, O, S, and Cl are converted to a syngas consisting of CO, H_2, H_2O, CO_2, NH_3, N_2, CH_4, H_2S, HCl, COS, HCN, elemental carbon, and traces of heavier hydrocarbon gases. The products of combustion processes are CO_2, H_2O, SO_2, NO, NO_2, and HCl.

Environmental Controls

Depending on the composition of feed material, combustion gases are processed in a series of process units to remove

TABLE 1.1 Comparison of Gasification and Combustion Technologies

Features	Gasification	Combustion
Purpose	Creation of valuable, usable products from waste or lower value material	Generation of heat or destruction of waste
Process Type	Thermal and chemical conversion using no or limited oxygen	Complete combustion using excess oxygen (air)
Raw Gas Composition (before gas cleanup)	H_2, CO, H_2S, NH_3, and particulates	CO_2, H_2O, SO_2, NO_x, and particulates
Gas Cleanup	Syngas cleanup at atmospheric to high pressures depending on the gasifier design	Flue gas cleanup at atmospheric pressure
	Treated syngas used for chemical, fuels, or power generation	Treated flue gas is discharged to atmosphere
	Recovers sulfur species in the fuel as sulfur or sulfuric acid	Any sulfur in the fuel is converted to SO_2 that must be removed using flue gas cleanup systems, generating a waste that must be landfilled.
	Clean syngas primarily consists of H_2 and CO.	Clean flue gas primarily consists of CO_2 and H_2O
Solid by-products/ products	Char or slag	Bottom ash
Ash/char or slag handling	Low temperature processes produce a char that can be sold as fuel.	Bottom ash and fly ash are collected, treated, and disposed as hazardous waste in most cases.
	High temperature processes produce a slag, a non-leachable, non-hazardous material suitable for use as construction materials.	

TABLE 1.1 (CONTINUED) Comparison of Gasification and
Combustion Technologies

Features	Gasification	Combustion
	Fine particulates are recycled to gasifier. In some cases fine particulates my be processed to recover valuable metals.	
Temperature	1300°F – 2700°F	1500°F – 1800°F
Pressure	Atmospheric to high	Atmospheric

particulates, heavy metals, and inorganic acid gases. These
process units may include gas cooling, followed by Venturi
scrubbers, wet electrostatic precipitators, or ionizing wet
scrubbers. Some facilities may also use packed tower absorbers for acid gas removal or fabric filters for particulate
removal. Demisters are usually used to remove visual vapor
before the combustion gases are emitted to atmosphere
through a stack. When the sulfur content of the fuel is high
and a very stringent SO_2 limit must be met, flue gas desulfurization processes such as sorbent injection in the combustor,
the flue gas duct, or added process units may have to be used
for sulfur removal. Sorbent addition not only adds to a combustion system's capital and operating costs but also increases
the amount of solid waste that must be landfilled if a suitable
end-use market cannot be identified for this byproduct.

Following the gasification step, the raw syngas is
quenched directly with water or cool recycled gas. Particulates
may also be removed using hot filters. Indirect cooling through
heat exchangers may follow syngas quenching before any
entrained particulates is removed. Syngas is then further
processed to remove sulfur compounds such as H_2S, COS, and
NH_3. Conventional treatment technologies, with sulfur
removal efficiencies of up to 99%, are utilized in the natural
gas and petroleum industries. The same conventional technologies can be used to recover sulfur as high-purity liquid
byproduct from raw syngas. When direct water quenching is

used, some particulates are captured by water and must be filtered. The particulate scrubber water and syngas condensates contain some water-soluble gases such as NH_3, HCN, HCl, and H_2S. These streams are usually recycled to the gasifier or scrubber after entrained solids are removed. A small portion of the water is purged from the system to prevent accumulation of dissolved salts. The purged water is then processed in a conventional wastewater treatment system.

Solid Byproducts

Solid by-products of gasification and combustion processes are significantly different. The primary solid by-product of a low-temperature gasification process is char. Char consists of unreacted carbon and the mineral matter present in the gasifier feed. The most important and significant use of char is as a source of activated carbon.

Char from a variety of sources, including coal, is used to produce activated carbon. The two most important uses for activated carbon are for water and wastewater treatment and decolorization. Other uses for activated carbon include the capture of pollutants such as volatile organic compounds (VOCs) and pesticide residues from industrial waste streams.

Other markets for char include iron, steel, and silicon/ferro-silicon industries. Char can be used as a reducing agent in direct reduction of iron. Ferro-silicon and metallurgical-grade silicon metal are produced carbothermally in electric furnaces. Silica is mixed with coke, either iron ore or scrap steel (in the case of ferro-silicon), and sawdust or charcoal in order to form a charge. The charge is then processed by the furnace to create the desired product. Char can be substituted for the coke as a source of reducing carbon for this process. Some plants in Norway are known to have used coal-char in the production of silicon-based metal products as late as mid-1990.[5] The use of char in this industry is not practiced due to lack of char supply.

---

[5]Rezaiyan, J., private communications with producers, 1997.

A solid byproduct of the high-temperature gasification process is slag, a glass-like material. It mainly consists of the inorganic materials in the gasifier feed that are not vaporized. Because of the gasifier's high temperature, above the fusion or melting temperature of the mineral matter, the mineral matter melts and is removed as molten slag, which forms a glassy substance upon quenching or cooling. The slag is usually found to be non-hazardous and can be used as an admix for road construction material or abrasive material for sand blasting. It can also be disposed of as non-hazardous waste. Depending on its composition it could also be sold for recovery of valuable metals.

The primary solid byproduct of combustion processes is bottom ash, which primarily consists of mineral matter and minor amounts of unreacted carbon. Because the leaching property of the ash, the bottom ash from combustion of most material is considered hazardous. An exception is the bottom ash from combustion of biomass.

Advantages of Gasification over Combustion

From an environmental standpoint, gasification offers several advantages over the combustion of solids, heavy oils, and carbonaceous industrial and domestic wastes. First, emission of sulfur and nitrogen oxides, precursors to acid rain, as well as particulates from gasification are reduced significantly due to the cleanup of syngas. Sulfur in the gasifier feed is converted to H_2S, and nitrogen in the feed is converted to diatomic nitrogen (N_2) and NH_3. Both H_2S and NH_3 are removed in downstream processes, producing a clean syngas. Therefore, if the resulting clean syngas is combusted in a gas turbine to generate electricity or in a boiler to produce steam or hot water, the production of sulfur and nitrogen oxides are reduced significantly. If the clean syngas is used as an intermediate product for manufacture of chemicals, these acid-rain precursors are not formed.

The particulates in the raw syngas is also significantly reduced due to multiple gas cleanup systems used to meet gas turbine manufacturers' specifications. Particulate removal takes place in primary cyclones, scrubbers, or dry filters and then in gas cooling and acid gas removal systems. One study by the U.S. Department of Energy (*The Wabash River Coal Gasification Repowering Project, Topical Report No. 7,* November 1996)[6] shows that repowering of conventional coal-fired utility systems with IGCC systems can reduce sulfur and nitrogen oxides as well as particulate emissions by one to two orders of magnitude.

A second major advantage is that furan and dioxin compounds are not formed during gasification. Combustion of organic matter is a major source of these highly toxic and carcinogenic pollutants. The reasons why furans and dioxins are not formed in gasification are:

1. The lack of oxygen in the reducing environment of the gasifier prevents formation of free chlorine from HCl and limits chlorination of any precursor compounds in the gasifier
2. High temperature of gasification processes effectively destroys any furan or dioxin precursors in the feed

Furthermore, if the syngas is combusted in a gas turbine where excess oxygen is present, the high combustion temperature does not favor formation of free chlorine. In addition, post-combustion formation of dioxin or furan is not expected to occur because very little of the particulates that are required for post-combustion formation of these compounds are present in the flue gas.

Dioxin or furan refers to molecules or compounds composed of carbon and oxygen. These compounds when reacted with halogens such as chlorine or bromine acquire toxic properties. Most research on halogenated dioxin and furan has been concerned with chlorinated species. It is generally accepted that dioxin and furan are by-products of combustion processes including domestic and medical waste combustion or incineration processes.[7] In combustion processes, hydrocar-

---

[6]See http://www.netl.doe.gov/cctc/factsheets/wabash/wabashrdemo.html
[7]See http://www.nihe.org/hcept_tool/dioxin.html.

bon precursors react with chlorinated compounds or molecules to form furan or dioxin. They may also form in a post-combustion flue gas cooling system due to due to presence of precursor compounds, free chlorine, or unburned carbon and copper species in the fly ash particles.[8]

Limited data is available on the concentration of volatile organic compounds, semi-volatile organic compounds (SVOCs), and polycyclic aromatic hydrocarbons (PAHs) from gasification processes. The data that is available indicate that VOCs, SVOCs, and PAHs are either non-detectable in flue gas streams from IGCC process or, in some cases where they were detected, they are at extremely low levels (on the order of parts per billion and lower). The analysis of syngas also indicates greater than 99.99 percent chlorobenzene and hexachlorobenzene destruction and removal efficiencies and part per billion or less concentration of selected PAHs and VOCs.[9–14]

STOICHIOMETRIES AND THERMODYNAMICS

As feedstock proceeds through a gasification reactor or gasifier, the following physical, chemical, and thermal processes

---

[8]Raghunathan, K. and Gullett, B. K., *Role of Sulfur in Reducing PCDD and PCDF Formation, Environmental Science and Technology*, Vol. 30. No. 6, pp. 1827-1834, June 1996.

[9]EPRI, *Summary Report: Trace Substance Emissions from a Coal-Fired Gasification Plant*, U.S. Department of Energy, June 1998.

[10]Baker, D. C., *Projected Emissions of Hazardous Air Pollutants from a Shell Gasification Process – Combined-Cycle Power Plant, Fuel*, Volume 73, No. 7, July 1994.

[11]Vick, S. C., *Slagging Gasification Injection Technology for Industrial Waste Elimination*, 1996 Gasification Technology Conference, San Francisco, CA, October 1996.

[12]Salinas, L., Bork, P., and Timm, E., *Gasification of Chlorinated Feeds*, 1999 Gasification Technologies Conference, San Francisco, CA, October 1999.

[13]DelGrego, G., *Experience with Low Volatile Feed Gasification at the El Dorado, Kansas Refinery*, 1999 Gasification Technologies Conference, San Francisco, CA, October 1999.

[14]U.S. EPA, *Texaco Gasification Process Innovation Technology Evaluation Report*, Office of Research and Development Superfund Innovation Technology Evaluation Program, EPA/540/R-94/514, July 1995.

may occur sequentially or simultaneously, depending on the reactor design and the feedstock material.

Drying

As the feedstock is heated and its temperature increases, water is the first constituent to evolve.

$$\text{Moist feedstock} + \text{Heat} \rightarrow \text{Dry feedstock} + H_2O$$

Devolatilization

As the temperature of the dry feedstock increases, pyrolysis takes place and the feedstock is converted to char.

$$\text{Dry feedstock} + \text{Heat} \rightarrow \text{Char} + \text{Volatiles}$$

Depending on the origin of the feedstock, the volatiles may include H_2O, H_2, N_2, O_2, CO_2, CO, CH_4, H_2S, NH_3, C_2H_6, and very low levels of unsaturated hydrocarbons such as acetylenes, olefins, and aromatics and tars. Char is the residual solids consisting of organic and inorganic materials. After pyrolysis, the char has a higher concentration of carbon than the dry feedstock.

Gasification

Gasification is the result of chemical reactions between carbon in the char and steam, carbon dioxide, and hydrogen in the gasifier vessel as well as chemical reactions between the resulting gases. Gasification reactions can be represented by:

$$C + H_2O + \text{Heat} \rightarrow CO + H_2$$

$$C + 2H_2O + \text{Heat} \rightarrow CO_2 + 2H_2$$

$$C + CO_2 + \text{Heat} \rightarrow 2CO$$

$$C + 2H_2 \rightarrow CH_4 + \text{Heat}$$

$$CO + 3H_2 + \text{Heat} \rightarrow CH_4 + H_2O$$

$$CO + H_2O + \text{Heat} \rightarrow H_2 + CO_2$$

Depending on the gasification process conditions, the remaining char may or may not have a significant amount of organic content or heating value.

Combustion

The thermal energy that derives gasification reactions must be provided directly, by combusting some of the char or dry feedstock and in some cases the volatiles within the gasifier, or indirectly, by combusting some of the feedstock, char, or clean syngas separately and transferring the required heat to the gasifier. The following chemical and thermal reactions my take place when char or dry feedstock is burned.

$$C + O_2 \rightarrow CO_2 + Heat$$

$$C + \tfrac{1}{2}O_2 \rightarrow CO + Heat$$

$$H_2 + \tfrac{1}{2}O_2 \rightarrow H_2O + Heat$$

$$Char + Heat \rightarrow Slag$$

$$Slag \rightarrow Clinker + Heat$$

Combustion of char or feedstock produces ash, unreacted organic material, which can be melted into liquid slag. Slag can be resolidified to form clinker.

In addition to heat, the combustion products are CO_2 and H_2O when clean syngas is burned to provide the required thermal energy.

It is not difficult to write a number of chemical equations to represent physical, thermal, and chemical reactions taking place in a gasification vessel. In theory, gasification processes can be designed so that heat release (exothermic reactions) balances the heat required by endothermic reactions. But in practice many of the above physical, thermal, and chemical reactions may take place simultaneously, making a precise prediction of the quantity and quality or composition of product gas somewhat difficult.

Thermodynamic and equilibrium characteristics of gasification systems, if available, could help to determine conditions under which certain desired products may be maximized. However, measurement of thermodynamic properties of feed material such as coal, char, biomass, and petroleum coke is very difficult due to the complex and heterogeneous nature of this material. Therefore, equilibrium characteristics of gasification systems are generally estimated using thermodynamic data (standard free energies of formation or standard enthalpies and entropies) for formation of pure reactants and products and simplified systems. The thermodynamic data for pure reactants and products of gasification systems can be found in a variety of tabulations and correlations.[15] Elliot,[16] Probstein and Hicks, and Klass present equilibrium composition of carbon-steam, carbon-oxygen-steam, carbon-hydrogen-oxygen, graphite (carbon)-hydrogen-methane, and graphite-carbon monoxide-carbon dioxide systems at different temperatures and pressures. The equilibrium data indicates that:

- CH_4 formation decreases with increasing temperature and increases with increasing pressures.
- CO and H_2 formation increases with increasing temperature and reducing pressures. Maximum concentration of H_2 and CO can be obtained at atmospheric pressure and temperature range of 800 to 1000°C.
- CO_2 concentration increases with increasing pressures and decreases sharply with increasing temperatures.
- Reducing oxygen-to-steam ratio of reactant gases (or reactor inlet streams) increases H_2 and CH_4 formation, while increasing the oxygen-to-steam ratio will increase CO and CO_2 formation.

Therefore, gasifier temperature and pressure can be controlled to maximize the concentration of desired product, CH_4 or H_2 and CO. However, in order to determine the optimum operating conditions other factors such as the gasification

[15] Elliot, M. A., Ed., *Chemistry of Coal Utilization,* 2nd Supplementary Volume, Wiley-Interscience, 1981.

[16] Probstein, R. F. and Hicks, R. E., *Synthetic Fuels,* McGraw-Hill, 1982.

Klass, D., *Biomass for Renewable Energy Fuels and Chemicals,* Academic Press, 1998.

kinetic, any catalyst effects, and the mechanism by which reactions occur must also be considered.

GASIFICATION KINETICS

An optimal design of a gasifier for converting a feed material to a desired syngas (i.e., low heating value, medium heating value, high heating value, high hydrogen concentration, high methane concentration) requires a detailed characterization of the relevant processes as well as a good understanding of physical and chemical processes that occur in a gasifier vessel. Most experimental studies have traditionally focused on coal gasification processes.[17-21]

More recently, researchers have also investigated biomass gasification.[22-26] Although most of these investigations assume a two-stage gasification process (a rapid pyrolysis stage followed by a slow char-hydrogen reaction), there is a considerable variation in the proposed mechanisms and kinetic representations used to correlate experimental data.

[17] Ergun, S. and Mentser, M., *Chemistry and Physics of Carbon*, P. L. Walker, Jr., Ed. Vol. 1, Marcel Dekker, New York, 1965.

[18] Blackwood, J. D. and McCarthy, D. J., *Aust. J. Chemical Engineering*, 19, 797-813 (1966).

[19] Blackwood, J. D. and McCarthy, D. J., *Aust. J. Chemical Engineering*, 20, 2,003-2,004 (1967).

[20] Mosely, F. and Paterson, D., *J. Inst. Fuel*, 40, 523-530 (1967).

[21] Johnson, J. L., *American Chemical Society*, Div. Fuel Chemistry, April 8, 1973.

[22] Satyanarayana, K. and Kearins, D. L., *Ind Eng. Chem. Fundamentals*, 20, 6-13, 1981.

[23] Dasappa, S., Paul, P. J., Mukunda, H. S., and Shrinivasa, U., *Wood-Char Gasification: Experiments and Analysis on Single Particles and Packed Beds*, 27th International Symposium on Combustion, The Combustion Institute, 1998, pp 1335-1342.

[24] See http://www.efpe.org/theses/Joaquin_Reina.pdf.

[25] See http://www.umsicht.fhg.de/WWW/UMSICHT/Produkte/ET/pdf/sevilla_v8_73-paper.pdf.

[26] Sadaka, S. S., Ghaly, A. E., and Sabbah, M. A., *Two Phase Biomass Air-Steam Gasification Model for Fluidized Bed Reactors: Part I – Model Development; Biomass and Bioenergy*, 22, 2002, pp. 439-462.

These variations are due to type of feed material used, the range of experimental conditions employed, the differences in the methods used to characterize experimental results, and the reactor system design.

In general, the first stage is considered to be instantaneous compared to the time required for char-hydrogen reactions. Most investigators characterize gasification processes using laboratory devices such as thermo-balance or muffles. However, the reaction conditions are very different in a fluidized bed- or entrained bed-type gasification reactor from those of laboratory devices. Practically, it is difficult to accurately measure particle temperature and mass as a function of time, flow, or composition of volatiles in a fluidized bed. However, there is general agreement among researchers that biomass compared to coal is more volatile, the biomass gasification process occurs under less severe conditions compared to coal gasification processes, and the char resulting from pyrolysis of biomass is more reactive than pyrolytic coal char.

The kinetics of gasification, particularly gasification of carbonaceous solids, is still the subject of intensive investigations and discussions; and existing gasification kinetics models are of limited value in designing commercial gasification reactors. Use of sophisticated computational tools[27] including probabilistic models[28] can help to develop gasification kinetic models that could be more useful in the design of gasification reactor systems in the feature. The basic kinetic theory is presented in this section to establish how gasification kinetics can be applied in the design of feature gasification systems.

Figure 1.2 shows a generally agreed sequence of gasification reactions for coal and biomass.[29] The gasification reaction

---

[27] See http://www.cis.tugraz.at/amft/science/pyrolyse/pyrolyse.en.html.

[28] See http://www.netl.doe.gov/publications/proceedings/03/ctua/posters/poster-eddings.pdf.

[29] Source: Adapted from various sources: M. A. Elliot, Ed., *Chemistry of Coal Utilization,* 2nd Supplementary Volume, Wiley-Interscience, 1981; Cheremisinoff, N, and P. Cheremisinoff, Hydrodynamics of Gas-Solid Fluidization, Gulf Publishing Co., 1984; Joaquin, R. H., *Kinetic and Hydrodynamic Study of Waste Wood Pyrolysis for its Gasification in Fluidized Bed Reactor,* European Foundation for Power Engineering, 2002 (http://www.efpe.org/papers.html).

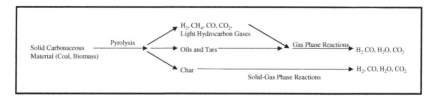

Figure 1.2   Coal gasification stages.

sequence for petroleum coke and heavy oils and tars are somewhat different. Petroleum coke has no or very little volatile content; heavy oils and tars consist primarily of volatile matter.

Pyrolysis processes were discussed in previous sections. Suffice it to say that if the heating rate is fast, a solid particle is heated to high temperatures in a short period of time, and then the gas-gas phase and solid-gas phase gasification reactions take place simultaneously.

If the heat rate is slow, then pyrolysis of solid particles begins at about 500 to 600°F. The gasification of volatiles and char is very slow at these temperatures. The concentration of volatiles increases until the gasifier is heated to temperatures of greater than 1100°F, at which steam gasification reactions are promoted. Depending on the reactor design configuration, pyrolysis gases may leave the reactor without significant gas-gas phase reactions taking place. Entrained and fluidized bed gasifiers exemplify fast heating-rate systems in which volatiles and char are gasified simultaneously, whereas counterflow fixed or moving bed gasifiers exemplify slow heating-rate gasifiers in which significant gas-gas phase gasification may not take place.

In air and oxygen blown gasifiers volatiles may also react with oxygen, producing CO_2, CO, and H_2O. When excess oxygen is available (e.g., in the combustion zone of a concurrent flow fixed-bed gasifier), the combustion of volatiles is complete. In the gasification zone or in a reducing environment this is not necessarily the case. Due to the heat and mass transfer limitations in the solid-gas phase reactions, the gas-gas phase reactions are much more rapid than the char-gas phase reactions. Therefore, char or carbon-gas reactions, or the heterogeneous gasification reactions, are the slowest reactions and govern the overall conversion reactions in coal and biomass gasification processes. These char-gas phase reactions are the Boudourd reaction (reaction 6), water-gas reactions (reactions 4 and 5), and hydro-gasification reaction (reaction 7), which is very slow except at high pressures, and methanation reaction (reaction 10), which is very slow relative to water-gas reactions unless catalyzed. Thus, one can assert that the predominant gasification reactions are the water-gas reactions:

$$(4) \ C + H_2O \rightarrow CO + H_2$$

$$(5) \ C + 2H_2O \rightarrow CO_2 + 2H_2$$

and the Boudourd reaction:

$$(6) \ C + 2H_2 \rightarrow CH_4.$$

BIOMASS GASIFICATION

Overview

When biomass is gasified, the producer gas is referred to as biogas. The objective of biomass gasification is to concentrate the energy content of biomass and convert it to a substance that is broadly usable in a variety of applications. Since the feedstock is most often based on agricultural feedstocks, gasification may be thought of as a renewable energy technology.

Facilities using renewable biomass to generate electricity currently produce enough power for about 7 million American households per year.[30] Many more biomass generation plants are not on the public power grid, but produce electricity and heat energy for manufacturing operations, primarily in the forest products industry.

The majority of these operations use boiler technology, which involves the direct combustion of biomass materials

---

[30] *Source:* http://www.eere.energy.gov/biopower/bplib/library/li_gasification.htm.

such as switchgrass, fast-growing trees, and wood waste — to produce steam to power electric generators. For grid-connected plants, competition from fossil fuels and the deregulation of the electric utility industry have resulted in the closing of some biomass power plants. In Europe, however, biomass gasification programs are being pursued more aggressively, with more emphasis given to small-scale plants for rural applications. It is estimated that large, commercial-scale gasifiers will use about 1,500 tons of biomass per day to generate up to 120 Megawatts of electricity, enough for about 120,000 households. Because it is a clean technology that uses renewable agricultural crops or manufacturing waste products as an energy source, gasification is ideal for community use and rural economic development.

Biomass gasifiers have the potential to be up to twice as efficient as using conventional boilers to generate electricity. For even greater efficiency, heat from the gas turbine exhaust can be used to generate additional electricity with a steam cycle. These improvements in efficiency can make environmentally clean biomass energy available at costs more competitive with fossil fuels.

Types of Biomass Gasifiers

Two major types of gasifiers currently in development are direct-fired gasifiers, using air, and the indirect-fired method, where heated sand surrounds biomass and gasifies it. These design schemes are described in Chapter 3.

Briefly, the leading direct-fired gasification process is the RENUGAS® system developed by the Institute of Gas Technology (IGT). This scheme uses air to produce a low-heating-value gas. A high-pressure fluidized-bed IGT gasification system is being demonstrated at a Hawaii Commercial and Sugar Company sugar processing facility on Maui, using sugar cane processing waste, known as bagasse, as a feedstock. While this plant currently produces electricity from agricultural residues, it is designed to accept a wide variety of biomass feedstocks.

The indirect-fired process gasifies biomass at low pressure, using indirect gasification to produce gases with medium heating values, that is, about half that of natural gas. Thus, little or no modifications are required to turbine combustors to burn the resulting gaseous fuels. A system based on the Battelle/Columbus indirect-fired biomass gasifier is nearing the demonstration phase at the McNeil Power Station in Burlington, VT.

In the Battelle/Columbus gasification system biomass particles are surrounded with extremely hot sand, which converts it into gaseous form. The solid biomass is surrounded by sand heated from 1800 to 1900°F, which converts the biomass into gas and residual char in a fluidized-bed reactor at 1500 to 1600°F. Sand is used to carry the biomass and the char and to distribute the heat. Using sand as a heat carrier keeps out the air. This results in a better quality fuel gas. A second reactor combusts the char to heat the sand. Remaining traces of condensable matter formed during gasification are removed in a chamber where a catalyst "cracks" and converts them into fuel gas. The clean biogas is then pressurized before it reaches the gas turbine.

Biomass Characteristics

The physical, chemical, and thermodynamic characteristics of biomass resources vary widely. This variation can occur among different samples of what would nominally seem to be the same resource. Also, variations could occur from one region to another, especially for waste products. This wide variation sometimes makes it difficult to identify a "typical" value to use when designing a gasification plant.

Table 1.2 provides some typical average values of the heating value for various biomass feedstocks. Selecting a single property value for design purposes can be problematic, and it is important to base designs on likely values and probable ranges. As an example, the density of suspended solid particles in raw sewage is reported to range from 100 to 350 mg/l,[31] while that of septic tank sludge is 310 to 93,378 mg/l.[32]

---

[31] Source: http://ohioline.osu.edu/aex-fact/0768.html.
[32] Source: EPA 832-f-99-068, September 1999.

TABLE 1.2   Typical Literature Reported Heating Values for
Various Biomass Sources

Biomass Source	Heating Value, Btu/Lb	Reference Source
Sewage sludge (biosolids)	8217	Klass, Donald L., *Biomass for Renewable Energy, Fuels, and Chemicals*, Academic Press, 1998.
Septage (biosolids)	8217	Klass, Donald L., *Biomass for Renewable Energy, Fuels, and Chemicals*, Academic Press, 1998.
Fruit pulp	3600	Assumed same as bagasse. Bagasse heating value obtained from Klass, Donald L. and G. H. Emert, *Fuels from Biomass and Waste*, Ann Arbor Science, 1981.
Wood waste (soft wood, firewood, construction waste)	87,133	Klass, Donald L., *Biomass for Renewable Energy, Fuels, and Chemicals*, Academic Press, 1998.
Mixed solid waste	4830	Encyclopedia of Chemical Technology, *Fuels from Waste*, Vol. 11.

Another example is that of Okara: the residue from the commercial processing of soybean contains 8% protein on wet
basis, or about 40% on a dry basis.[33] Thus, the wet Okara
contains about 80% moisture.

Wood and wood waste sources can come from softwood
or hardwood, and there are significant differences in the heating
values, as well as physical and chemical properties. Because
the moisture content of green biomass can be quite high and
can negatively impact the conversion of biomass to energy
processes, pre-drying may be needed. Moisture content of 10
to 20% is usually preferred. The construction lumber is generally kiln-dried to ensure uniform moisture among different
pieces. Dry lumber, as defined in the American Softwood Lumber Standard, has maximum moisture of 19%.[34] However,

---

[33] Source: http://www.ag.uiuc.edu/~intsoy/soymilk.htm.
[34] *Wood Handbook*, Forest Products Laboratory, 1999.

wood exposed to outdoor atmosphere reaches a moisture equilibrium content depending on the humidity and temperature.

Uncertainties with the availability and suitability of biomass resources for energy production are primarily due to their varying moisture content, and to a lesser degree to their chemical composition and heating value. As the moisture content of biomass increases, the efficiency of thermal conversion process decreases. At some point more energy may have to be expended to dry the biomass than it contains. Uncertainties can be reduced by conducting a detailed chemical and physical analysis of the biomass sources.

PETROLEUM COKE GASIFICATION

Refiners are being pushed towards producing cleaner, lower sulfur transportation fuels from poorer quality crudes. Petroleum coke (pet coke) could be used as a source of hydrogen. Hydrogen will be in great demand as the Tier 2 sulfur regulations limiting sulfur in gasoline to 30 ppm and sulfur in diesel to 15 ppm take effect. Gray and Tomlinson[35] have noted that pet coke can be converted via gasification into clean synthesis gas, and liquid products can be made by application of Fisher-Tropsch technology. F-T liquids are zero sulfur, paraffinic hydrocarbons that can be classified as ultra-clean transportation fuels. Zero sulfur, high cetane F-T diesel could be used as a blending stock to assist refiners in meeting ultra low sulfur diesel specifications. In addition, gasified pet coke could be used to produce refinery power, and excess power could be sold. In a deregulated electric power industry, refiners may choose to become power providers.

Published oil industry data[36] show that there are 35 U.S. refineries producing more than 1000 tons per day of pet coke. A total of almost 95,000 tons per day of petroleum coke is

---

[35] Gray, D. and Tomlinson, G., "Opportunities for Petroleum Coke Gasification Under Tighter Sulfur Limits for Transportation Fuels," Paper presented at the 2000 Gasification Technologies Conference, San Francisco, CA, October 8–11, 2000.

[36] *Oil and Gas Journal*, December 20, 1999.

produced in these 35 refineries. Total U.S. coke production for that year was 96,200 tons therefore these refineries represent over 98% of U.S. production. Based on total crude capacity this production of coke is equivalent to 12.5 tons per thousand barrels feed per day. The actual feed to the cokers was 1.6 million barrels per day (MMBPD), to give an average coke yield of about 57 tons per thousand barrels feed.

Assuming that demand for petroleum continues to increase at a rate of 1.2% per annum to 2010,[37] and that all gasoline and diesel produced by U.S. refineries will have a sulfur content of less than 30 ppm, desulfurization of gasoline and diesel to these low levels will require extensive hydrotreating of both catalytic cracker feed and product of distillate.

Gray and Tomlinson have applied a refinery simulation model that estimates the hydrogen required and costs of this desulfurization. The results of this model show that an average 150,000 barrel per day (BPD) refinery will require an additional 38 MMSCFD of hydrogen to produce gasoline and diesel with a sulfur content of less than 30 ppm. This is equivalent 0.25 MMSCFD per 1000 BPD. Their study investigated the potential for petroleum coke conversion at a generic refinery to produce a combination of products including hydrogen, electric power, and ultra-clean F-T liquid fuels. Their analysis supports that pet coke could be a candidate feedstock for hydrogen production if refiners have to pay in excess of $3.25/MMBtu for natural gas, as steam reformer feed and oil prices stay above $25 per barrel.

Hydrogen availability is an important issue and refiners must be persuaded that gasification will prove to be as reliable a technology in the future as natural gas steam reforming is today. Many refineries produce sufficient pet coke to more than satisfy refinery hydrogen requirements. This would allow co-production of hydrogen and power or F-T liquids.

---

[37] Energy Information Administration Annual Energy Outlook 1999 With projections to 2020, DOE/EIA-0383(99), December 1998.

Advances in established refining technologies have enhanced options for economically processing and using residues.

Petroleum coke can be used as either a primary or a secondary fuel in a new grassroots plant or for co-firing in an existing coal-fired power plant. A large percentage of petroleum coke used in power generation in the U.S. is for co-firing in existing suspension boilers. Because coke has superior heating value and negligible ash content, it is typically blended with coal. Blending has a positive impact on reducing operation and maintenance (O&M) costs. Because of the low volatile matter of petroleum coke, the blending ratio is generally kept below 20%. This is done to ensure stable flame and preclude ignition problems. Another reason for maintaining a low blending ratio is that the high level of sulfur content in petroleum coke could necessitate the addition of flue gas desulfurization (FGD) equipment to remain within the allowable emission limits. At least 15 U.S. utilities are currently using petroleum coke for co-firing in existing boilers.

The Ube Ammonia plant in Japan is the oldest commercial gasification unit operating with coke. Although originally designed for coal gasification, the attractive pricing of coke in Japan resulted in a gradual change in feedstock. In 1996 Texaco started up a coke gasification facility at Texaco's El Dorado refinery near Wichita, KS. The gasification facility is designed to supply one third of the fuel needs of the refinery's cogeneration plant of 35 MW and 82,000 kg/hr process steam. In mid-1997 the power island of Elcogas's 300 MW gasification-combined cycle unit in Puertolano, Spain, came on line, initially using only natural gas. The first firing of the gasifier was accomplished in December 1997. This facility is designed to use a feedstock of 50% coal and 50% coke. Another CGCC unit on order in the U.S. is Star Refinery's nominal 180 MW (excluding existing steam turbine) Delaware City repowering facility with 295,000 kg/hr of 41 bar process steam. As a large number of other units in the 250 MW range that use feedstocks ranging from coal to refinery residues have recently come on line or are in the design stages, the gasification technology is coming of age.

REFERENCES

1. Lowry, H. H., editor, *Chemistry of Coal Utilization,* John Wiley & Sons, 1945, p. 1252.

2. Cyclic gas generators converted coke, a by-product of high-temperature pyrolysis process, to a synthetic gas by alternatively exposing the coke to air to provide heat and to steam to produce a gas that burned with a blue flame. The coal gas was know as "blue water gas" (Probstein, R. F. and Hicks, R. E., *Synthetic Fuels,* McGraw-Hill, 1982, p. 7).

3. Klass, D. L., *Biomass for Renewable Energy, Fuels, and Chemicals,* Academic Press, 1998, p. 271.

4. U.S. Department of Energy, http://www.netl.doe.gov/coalpower/gasification/models.

5. Rezaiyan, J., private communications with producers, 1997.

6. See http://www.netl.doe.gov/cctc/factsheets/wabash/wabashrdemo.html.

7. See http://www.nihe.org/hcept_tool/dioxin.html.

8. Raghunathan, K. and Gullett, B. K., *Role of Sulfur in Reducing PCDD and PCDF Formation, Environmental Science and Technology,* Vol. 30. No. 6, pp. 1827–1834, June 1996.

9. EPRI, *Summary Report: Trace Substance Emissions from a Coal-Fired Gasification Plant,* U.S. Department of Energy, June 1998.

10. Baker, D. C., *Projected Emissions of Hazardous Air Pollutants from a Shell Gasification Process – Combined-Cycle Power Plant, Fuel,* Volume 73, No. 7, July 1994.

11. Vick, S. C., *Slagging Gasification Injection Technology for Industrial Waste Elimination,* 1996 Gasification Technology Conference, San Francisco, CA, October 1996.

12. Salinas, L., Bork, P., and Timm, E., *Gasification of Chlorinated Feeds,* 1999 Gasification Technologies Conference, San Francisco, CA, October 1999.

13. DelGrego, G., *Experience with Low Volatile Feed Gasification at the El Dorado, Kansas Refinery,* 1999 Gasification Technologies Conference, San Francisco, CA, October 1999.

14. U.S. EPA, *Texaco Gasification Process Innovation Technology Evaluation Report,* Office of Research and Development Superfund Innovation Technology Evaluation Program, EPA/540/R-94/514, July 1995.

15. Elliot, M. A., Ed., *Chemistry of Coal Utilization,* 2nd Supplementary Volume, Wiley-Interscience, 1981.

16. Probstein, R. F. and Hicks, R. E., *Synthetic Fuels,* McGraw-Hill, 1982. Klass, D., *Biomass for Renewable Energy Fuels and Chemicals,* Academic Press, 1998.

17. Ergun, S. and Mentser, M., *Chemistry and Physics of Carbon,* P. L. Walker, Jr., Ed. Vol. 1, Marcel Dekker, New York, 1965.

18. Blackwood, J. D. and McCarthy, D. J., *Aust. J. Chemical Engineering,* 19, 797-813 (1966).

19. Blackwood, J. D. and McCarthy, D. J., *Aust. J. Chemical Engineering,* 20, 2003–2004 (1967).

20. Mosely, F. and Paterson, D., *J. Inst. Fuel,* 40, 523–530 (1967).

21. Johnson, J. L., *American Chemical Society,* Div. Fuel Chemistry, April 8, 1973.

22. Satyanarayana, K. and Kearins, D. L., *Ind Eng. Chem. Fundamentals,* 20, 6–13, 1981.

23. Dasappa, S., Paul, P. J., Mukunda, H. S., and Shrinivasa, U., *Wood-Char Gasification: Experiments and Analysis on Single Particles and Packed Beds,* 27th International Symposium on Combustion, The Combustion Institute, 1998, pp 1335–1342.

24. See http://www.efpe.org/theses/Joaquin_Reina.pdf.

25. See http://www.umsicht.fhg.de/WWW/UMSICHT/Produkte/ET/pdf/sevilla_v8_73-paper.pdf.

26. Sadaka, S. S., Ghaly, A. E., and Sabbah, M. A., *Two Phase Biomass Air-Steam Gasification Model for Fluidized Bed Reactors: Part I – Model Development; Biomass and Bioenergy,* 22, 2002, pp. 439–462.

27. See http://www.cis.tugraz.at/amft/science/pyrolyse/pyrolyse.en.html.

28. See http://www.netl.doe.gov/publications/proceedings/03/ctua/posters/poster-eddings.pdf.

29. Source: Adapted from various sources: M. A. Elliot, Ed., *Chemistry of Coal Utilization,* 2nd Supplementary Volume, Wiley-Interscience, 1981; Cheremisinoff, N, and P. Cheremisinoff, Hydrodynamics of Gas-Solid Fluidization, Gulf Publishing Co., 1984; Joaquin, R. H., *Kinetic and Hydrodynamic Study of Waste Wood Pyrolysis for its Gasification in Fluidized Bed Reactor,* European Foundation for Power Engineering, 2002 (http://www.efpe.org/papers.html).

30. Source: http://www.eere.energy.gov/biopower/bplib/library/li_gasification.htm.

31. Source: http://ohioline.osu.edu/aex-fact/0768.html.

32. Source: EPA 832-f-99-068, September 1999.

33. Source: http://www.ag.uiuc.edu/~intsoy/soymilk.htm.

34. *Wood Handbook,* Forest Products Laboratory, 1999.

35. Gray, D. and Tomlinson, G., "Opportunities for Petroleum Coke Gasification Under Tighter Sulfur Limits for Transportation Fuels," Paper presented at the 2000 Gasification Technologies Conference, San Francisco, CA, October 8–11, 2000.

36. *Oil and Gas Journal*, December 20, 1999.

37. Energy Information Administration Annual Energy Outlook 1999 With projections to 2020, DOE/EIA-0383(99), December 1998.

RECOMMENDED RESOURCES

1. Zierold, D., Voyles, R., and Casada, M., "NISCO: FBC Unit Use of Petroleum Coke." EPRI Workshop, Ft. Lauderdale, FL, May 1993.

2. Dickenson, R. L., Biasen, F. E., Schulman, B. L., and Johnson, H. E., "Refinery Options for Converting and Utilizing Heavy Fuel Oil." *Hydrocarbon Processing*, February 1997.

3. Swain, E. "Coke, Sulfur Recovery from U.S. Refineries Continues to Increase." *Oil & Gas Journal*, January 2, 1995.

4. Holt, N. "Petroleum Coke Utilization for Power Generation." EPRI Petroleum Coke Workshop, Ft. Lauderdale, FL, May 1993.

5. Geosits, R. and Mohammad-Zadeh, Y., "Coke Gasification Power Generation: Options and Economics." *Power-Gen Americas*, Dallas, TX, November 1993.

6. Pohani, B. and Wen, H., "Comparison of Circulating Fluidized Bed Combustion and Pulverized Combustion for Coke Fired Cogeneration Plant." Ninth Annual Energy-Sources Technology Conference and Exhibition, New Orleans, LA, February 1986.

7. O'Conner, D. "Direct Firing of Petroleum Coke: Utility Experience and Issues." EPRI Petroleum Coke Workshop, May 1993.

8. "Co-firing pet coke: Two steps forward, one back." *Power*, September/October 1996.

9. Rossi, R. A. "Refinery byproduct emerges as a viable power plant fuel." *Power*, August 1993.

10. Genereux, R. P. and Doucette, B., "Pet-coke-firing experience evolves over three decades." *Power*, July/August 1996.

11. Showyra, R. "Power Generation Requirements." Eleventh Annual Fluidized Bed Conference, Allentown, PA, November 1995.

12. Jones, C. "O&M experience underscores maturity of CFB technology." *Power*, May 1995.

13. Jeffs, E. "Karita: A quantum leap for PFBC." *Turbomachinery International*, March/April 1997.

14. Jahanke, F. C., Falsetti, J. S., and Wilson, R. F., "Coke Gasification, Costs, Economics & Commercial Applications." 1996 NPRA Annual Meeting, San Antonio, TX, March 1996.

15. Tavoulareas, E. S., and Charpentier, J. P., "Clean Coal Technologies for Developing Countries." World Bank Technical Paper Number 286.

16. Hennagir, T. "PFBC Progresses." *Independent Energy*, September 1996.

17. Refining Processes '96. *Hydrocarbon Processing*, November 1996.

18. Chang, T. "Worldwide Refining." *Oil & Gas Journal*, December 22, 1997.

2

Coal Gasification Technologies

INTRODUCTION

This chapter provides an overview of the leading commercial and semi-commercial technologies suitable for coal gasification.

Gasification can be thought of as a combustion process in which sufficient air is supplied to allow a portion of the carbonaceous feed material to burn. The heat that is generated is used to de-volatize and decompose the majority of the remaining feed material into hydrocarbon gases.

The gas stream produced is a mixture of the inert flue gases and hydrocarbons. This producer gas, or syngas, has a calorific value (CV). The gas stream generally contains a large portion of nitrogen, which can be up to 60%, as a result of using air.

Some processes employ oxygen or steam to provide the necessary oxygen. These systems produce a gas stream with a much higher CV. However, this introduces both additional costs and safety issues.

Gasification can be carried out in many different reactor types including:

- Fixed Bed
- Fluid Bed
- Bubbling, Circulating, Entrained, Twin Bed

- Moving Bed
- Rotary Kiln
- Cyclonic

Each of these technologies are discussed to the point of providing a working knowledge of the process, along with the advantages and disadvantages of each. Before discussing the processes, the chapter begins with an overview of coal gasification and the properties of coal that should be considered when evaluating the suitability of gasification for an application.

COAL GASIFICATION

Overview

Coal gasification is a process that converts coal from a solid to a gaseous fuel through partial oxidation. Once the fuel is in the gaseous state, undesirable substances, such as sulfur compounds and coal ash, may be removed from the gas. The net result is a clean, transportable gaseous energy source.

In contrast to combustion process, which works with excess air, the gasification process works on partial combustion of coal with the oxygen supply controlled (generally 20 to 70% of the amount of O_2 theoretically required for complete combustion) such that both heat and a new gaseous fuel are produced as the coal is consumed. In simplest terms, the stoichiometric reactions are as follows:

$$C + \frac{1}{2}O_2 \text{ gasification} \rightarrow CO$$

$$C + H_2O \text{ gasification} \rightarrow CO + H_2$$

Types of Coal

There are several different types of coal, each displaying different properties resulting from their age and the depth to which they have been buried under other rocks. In some parts of the world (e.g., New Zealand), coal development is accelerated by volcanic heat or crustal stresses.

The degree of coal development is referred to as a coal's "rank," with peat being the lowest rank coal and anthracite the highest. The various types are as follows:

Peat. Peat is the layer of vegetable material directly underlying the growing zone of a coal-forming environment. The vegetable material shows very little alteration and contains the roots of living plants. Peat is widely used as a domestic fuel in rural parts of the world.

Lignite. Lignite is geologically very young (upward of around 40,000 years). It is brown and can be soft and fibrous, containing discernible plant material. It also contains large amounts of moisture (typically around 70%) and hence, it has a low energy content (around 8 to 10 MJ/kg). As the coal develops it loses its fibrous character and darkens in color.

Black coal. Black coal ranges from Cretaceous age (65 to 105 million years ago) to mid-Permian age (up to 260 million years ago). They are all black; some are sooty and still quite high in moisture (sub-bituminous coal). A common name for this coal in many parts of the world is "black lignite." Coals that get more deeply buried by other rocks lose more moisture and start to lose their oxygen and hydrogen; they are harder and shinier (e.g., bituminous coal). Typical energy contents are around 24 to 28 MJ/kg. These coals generally have less than 3% moisture, but some power stations burn coal at up to 30% ash.

Anthracite. Anthracite is a hard, black, shiny form of coal that contains virtually no moisture and very low volatile content. Because of this, it burns with little or no smoke and is sold as a "smokeless fuel." In general, coals only approach anthracite composition where bituminous coal seams have been compressed further by local crustal movements. Anthracites can have energy contents up to about 32 MJ/kg, depending on the ash content.

It is important to note that coal rank has little to do with quality. As a coal matures its ash content actually increases as a proportion because of the loss of moisture and volatiles. Lower rank coals may have lower energy contents, but they tend to be more reactive (i.e., they burn faster) because of their porosity and resultant higher surface area.

Composition and Structure

Coal is an organic deposit that is made up of organic grains called macerals. Macerals are categorized into three groups, each of which are composed of several maceral types. These groups are liptinite, vitrinite, and inertinite. The groups are defined according to their grayness in reflected light: liptinites are dark gray, vitrinites are medium to light gray, and inertinites are white and can be very bright.

Liptinites were made up of hydrogen-rich hydrocarbons derived from spores, pollens, cuticles, and resins in the original plant material. Vitrinites were made up of wood, bark, and roots and contained less hydrogen than the liptinites. Inertinites are mainly oxidation products of the other macerals and are consequently richer in carbon. The inertinite group includes fusinite, most of which is fossil charcoal, derived from ancient peat fires.

Coals are also divided into two types based on their macroscopic appearance: banded and non-banded. Non-banded coals include cannel and boghead coals, both of which are dull and blocky.

Cannel is derived from the word "candle," because pencil-shaped pieces were used as candles in the past. Banded coals grade from dull banded ("splint coal") to bright banded coals, depending upon whether dull bands or bright bands are dominant. The bands are divided into lithotypes. Dull bands are called durain; satiny bands are clarain; charcoal bands are fusain; and black, glassy bands are vitrain. Bright coals have lots of vitrain and clarain; dull coals are rich in durain bands. Fusain generally occurs only in thin and sporadic bands.

Splint coals are durain-rich and can be massive (non-banded) or banded. Most vitrain- and clarain-rich banded coals break into small blocky pieces along joints called cleats. Vitrain and clarain are brittle and break easily.

"Block coals" are dull coals that break into large blocks because they have fewer vitrain and clarain bands, but have a composition higher in liptinite macerals, which are tough.

"Bone" and "bone coals" have a high ash content in the form of clays and silts; they form part of a continuum between dark shale and dull (banded or non-banded) coal in the following sequence: dark shale, bone (greater than 50% ash), boney coal (less than 50%), dull coal (cannel, boghead, or splint).

Characteristics

The characteristics of coals that determine classification and suitability for given applications are the proportions of volatile matter, fixed carbon, moisture, sulfur, and ash.

Each of these is reported in the proximate analysis. Coal analyses can be reported on several bases: as-received, moisture-free (or dry), and mineral-matter-free (or ash-free).

As-received is applicable for combustion calculations; moisture-free and mineral-matter-free, for classification purposes.

Volatile matter is driven off as gas or vapor when the coal is heated according to a standard temperature test. It consists of a variety of organic gases, generally resulting from distillation and decomposition. Volatile products given off by coals when heated differ materially in the ratios (by mass) of the gases to oils and tars. No heavy oils or tars are given off by anthracite, and very small quantities are given off by semianthracite.

As volatile matter in the coal increases to as much as 40% of the coal (dry and ash-free basis), increasing amounts of oils and tars are released. However, for coals of higher volatile content, the quantity of oils and tars decreases and is relatively low in the sub-bituminous coals and in lignite.

Fixed carbon is the combustible residue left after the volatile matter is driven off. It is not all carbon. Its form and hardness are an indication of fuel coking properties and, therefore, serve as a guide in the selection of combustion equipment. Generally, fixed carbon represents that portion of fuel that must be burned in the solid state.

Moisture is difficult to determine accurately because a sample can lose moisture on exposure to the atmosphere, particularly when reducing the sample size for analysis. To correct for this loss, total moisture content of a sample is customarily determined by adding the moisture loss obtained when air-drying the sample to the measured moisture content of the dried sample. Moisture does not represent all of the water present in coal; water of decomposition (combined water) and of hydration are not given off under standardized test conditions.

Ash is the noncombustible residue remaining after complete coal combustion. Generally, the mass of ash is slightly less than that of mineral matter before burning. Sulfur is an undesirable constituent in coal, because the sulfur oxides formed when it burns contribute to air pollution and cause combustion system corrosion.

Heating value may be reported on an as-received, dry, dry and mineral-matter-free, or moist and mineral-matter-free basis. Higher heating values of coals are frequently reported with their proximate analysis. When more specific data are lacking, the higher heating value of higher-quality coals can be calculated by the Dulong formula:

Higher heating value, Btu/lb =

$$\left(14{,}544C + 62{,}028\left[H - (O/8)\right]\right) + 4050S \qquad (2.1)$$

where C, H, O, and S are the mass fractions of carbon, hydrogen, oxygen, and sulfur in the coal, respectively.

Literature reported values for ultimate analyses for coals are reported in Table 2.1.

GASIFIER CONFIGURATIONS

Gasifier Classification

Coal gasification technologies can be classified according to the flow configuration of the gasifier unit. The primary configurations are:

- Entrained flow
- Fluidized bed
- Moving bed

TABLE 2.1   Ultimate Analyses for Coals (miscellaneous Web sites)

Rank	As Received, Btu/Lb	Percentage by Mass					
		O	H	C	N	S	Ash
Anthracite	12,700	5.0	2.9	80.0	0.9	0.7	10.5
Semianthracite	13,600	5.0	3.9	80.4	1.1	1.1	8.5
Low-volatile bituminous	14,350	5.0	4.7	81.7	9.4	1.2	6.0
High-volatile bituminous A	13,800	9.3	5.3	75.9	1.5	1.5	6.5
High-volatile bituminous B	12,500	13.8	5.5	67.8	1.4	3.0	8.5
High-volatile bituminous C	11,000	20.6	5.8	59.6	1.1	3.5	9.4
Sub-bituminous B	9,000	29.5	6.2	52.5	1.0	1.0	9.8
Lignite	6,900	44.0	6.9	40.1	0.7	1.0	7.3

Figure 2.1 provides listing of the leading gasification technology developer/supplier by technology classification.

Typical operating characteristics of coal gasifiers are as follows:

	Moving Bed	Fluidized Bed	Entrained Bed
Exit Gas Temp. °C	420–650	920–1050	1200
Coal Feed Size	<50 mm	<6 mm	<100 MESH
Ash Conditions	Dry/Slagging	Dry/Agglomerating	Slagging

The following provides brief descriptions of the key features of each technology listed in Figure 2.1.

ENTRAINED FLOW TECHNOLOGIES

***Hitachi*:** This coal gasification technology is based on an oxygen-blown entrained flow gasifier, where the majority of experience has been gained in a 150 ton coal/day unit. The gasifier is a water-cooled tube that is lined by a high-temperature-resistant castable. Pulverized coal is pneumatically

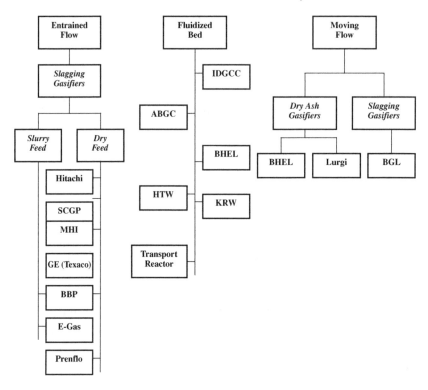

Figure 2.1   Coal gasification technology suppliers by technology classification by flow geometry.

transported by nitrogen to the gasifier, where it is injected into the gasifier chamber through two types of burners at a pressure of 2.5 MPa. The two sets of burners are installed tangentially to the gasifier sidewall, allowing a spiral flow of coal and oxygen from the upper stage to the lower stage and making particle residence times much longer than those of a gas stream. Enough oxygen is fed to the lower burner to melt the slag. Molten slag solidifies on the gasifier wall as a first layer, and subsequent molten slag flows over the layer of the solidified slag to the slag tap hole at the bottom of the gasifier; it is quenched with water and finally removed via a lock hopper. Coal fed to the upper burners is reacted at a lower temperature with a smaller amount of oxygen; it is then gasified and converted to reactive char. The char moves down along the spiral gas flow and mixes with high-temperature gas in the lower portion of the gasifier, where gasification proceeds further. The raw gas produced together with the fly ash and the remaining char particles go up toward the exit of the gasifier. They enter a syngas cooler, where they are cooled to 450°C prior to going through a cyclone and a filter that retain most of the fly ash and the char particles, which are finally reinjected into the gasifier by pneumatic transport under nitrogen. The syngas goes successively through a water scrubber to remove halides and is desulfurized to be cleaned enough to comply with the strict tolerance limits of fuel cells.

SCGP: The Shell Coal Gasification Process (SCGP) can operate on a wide variety of feedstocks. It consists of three principal stages:

1. Gasification (Partial Oxidation), in which the feedstock is converted to syngas in the presence of oxygen and a moderating agent (steam) in a refractory-lined gasification reactor
2. Syngas Effluent Cooler (SEC), in which high-pressure steam is generated from the hot syngas leaving the reactor
3. Carbon Removal, in which residual carbon and ash are removed from the syngas in a two-stage water scrubbing unit

The Shell gasifier is a dry-feed, pressurized, entrained slagging gasifier. Figure 2.2 illustrates the basic process. Feed coal is pulverized and dried with the same type of equipment used for conventional pulverized coal boilers. The coal is then pressurized in lock hoppers and fed into the gasifier with a transport gas by dense-phase conveying. The transport gas is usually nitrogen; however, product gas can be used for synthesis gas chemical applications, where nitrogen in the product gas is undesirable. The oxidant is preheated to minimize oxygen consumption and mixed with steam as a moderator prior to feeding to the burner. The coal reacts with oxygen at temperatures in excess of 2500°F to produce principally hydrogen and carbon monoxide with little carbon dioxide.

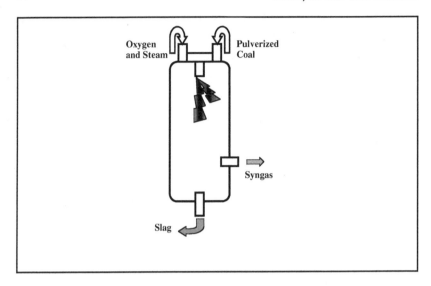

Figure 2.2   The Shell gasifier.

Operation at elevated temperatures eliminates the production
of hydrocarbon gases and liquids in the product gas. The high-
temperature gasification process converts the ash into molten
slag, which runs down the refractory-lined water wall of the
gasifier into a water bath, where it solidifies and is removed
through a lock hopper as a slurry in water. Some of the molten
slag collects on the cooled walls of the gasifier to form a
solidified protective coating. The crude raw gas leaving the
gasifier at 2500 to 3000°F contains a small quantity of unburned
carbon and about half of the molten ash. To make the ash
non-sticky, the hot gas leaving the reactor is partially cooled
by quenching with cooled recycle product gas. Further cooling
takes place in the waste heat recovery (syngas cooler) unit,
which consists of radiant, superheating, convection, and econ-
omizing sections, where high-pressure superheated steam is
generated before particle removal.

The first commercial IGCC plant using the Shell Coal
Gasification Process (SCGP) is Buggenum in the Netherlands,
which was built in 1993. The plant achieves an overall efficiency of 43%, which could be increased to over 50% if using

the most recent gas turbines. The Buggenum design processes coal with natural gas as backup. The plant can process up to 2000 ton/day of fuel. A demonstration plant (220 ton/day) at Oil Deer Park Manufacturing complex in Houston completed tests proving the ability of the SCGP to gasify more diverse types of coals (220 ton/day of bituminous coals or 365 ton/day of high-moisture, high-ash lignite) before being shut down in 1991. Any coal that can be milled to the right size and pneumatically transported can be gasified in the Shell entrained flow gasifier. Some adjustments have to be made in order to keep the SCGP performances optimal when changing coal. Bituminous coals require, in most cases, steam injection and oxygen/MAF (moisture- and ash-free) coal ratios from 0.85 to 1.05 for producing a syngas with a CO/H_2 ratio of 2.2 to 2.4 and 1 to 2.5% CO_2. Sub-bituminous coals and lignites normally do not require steam injection and can be operated with oxygen/MAT coal ratio between 0.8 and 0.9, producing syngas with some 3 to 5% CO_2 and a CO/H_2 ratio of 2.0 to 2.2. Anthracites require a higher oxygen/MAF coal ratio of 1.0 to 1.1 and a higher steam/oxygen ratio of 0.15 to 0.3, and they produce a syngas with similar CO_2 contents as bituminous coal (1 to 2.5% CO_2, but a higher CO/H_2 ratio of 2.4 to 2.6). The ash content of a coal has an impact on the performance of the SCGP process in terms of efficiency, as slag forms part of the insulation of the wall of the gasifier and then prevents excessive heat loss during the gasification reaction.

A new Integrated Gasification Combined Cycle (IGCC) project based on the SCGP technology, was also proposed to be built at Sulcis in Sardinia, Italy. It was planned to have similar characteristics as the Buggenum plant. The Sulcis plant has been designed to gasify 5000 ton/day blends of local coal (high-sulfur, high-ash sub-bituminous coal) and imported LHV coals. A large IGCC demonstration plant is also planned to be built at Yantai Power plant in Shandong province in China. Technical prefeasibility studies were carried out in 1994–95. Development prospects were predicted and comparisons were made with CFBC, PFBC-CC, and supercritical units. Two 400 MW IGCC units were proposed to be installed. Their net efficiency is planned to be more than 43%. They are designed to gasify bituminous coals with high sulfur content (2.5 to 3%) from Yanzhou in Shangong. Sulfur will be recovered as elemental sulfur with a predicted removal efficiency of 98%. Three other gasification plants are planned to be developed by Shell in partnership with Sinopec in China, and a fourth one is under feasibility study. The plants will all produce syngas for ammonia/urea production or H_2 for other chemical plants (methanol, oxo), replacing naphta reformers, oil gasifiers, or outdated coal gasifiers.

MHI: Mitsubishi Heavy Industries (MHI) consists of an air-blown MHI gasifier that is divided into two sections: a lower combustion section, which is connected by a diffuser to an upper reducing section. Dry pulverized coal is fed at two points into the gasifier, with half of the coal being fed into the combustor together with air, where it is burned to produce CO. The temperature inside the combustor is sufficiently high to melt the coal ash without the addition of flux. The slag runs to the bottom of the gasifier, where it is quenched in a water bath and removed using a lock hopper system. The gas produced in the combustor rises to the reducing section, where the remaining coal is added. Coal is then gasified in the reducing section to produce a low CV syngas mainly formed of nitrogen. As the reducer section is at a lower temperature than the combustor section, any molten ash carried upwards is solidified. The syngas produced exits the gasifier through a syngas cooler, and then cyclones are used to collect the chars as the coal is not completely gasified in the reducing section. Chars collected in the cyclones are then reinjected at the base of the gasifier to ensure complete carbon conversion. Because of the very high temperatures reached in the combustion section, this type of gasifier is well suited to gasify the very high ash-melting point Australian coals without any addition of fluxing agent.

The MHI gasification technology has been tested in Nakoso (Japan) in two pilot-scale gasifiers. A new IGCC project has been started that is a 250 MW air-blown IGCC demonstration plant located in Nakoso, where the former pilot plants were based, and will process up to 1500 ton/day of coal, which is about nine times more than the former 200 ton/day pilot plant. The system will have a unique feature: The oxidizing gas will be partially extracted from the gas turbine compressor and will be enriched with oxygen coming from an independent air separation unit, making the gasifier operation more stable and giving a certain flexibility to the system that does not exist in the two highly integrated European IGCC plants. An advantage of the MHI two-stage dry-fed entrained flow gasifier is that the syngas temperature at the outlet of the gasifier is not as high as the one flowing out of a one-stage gasifier. This means that the process does not require a large radiant cooler or a quenching system to mix cold recycled gas with the syngas. The overall cost of the process should then be less than that of existing IGCC plants. The raw gas produced together with the fly ash and the remaining char particles go up toward the exit of the gasifier. They enter a syngas cooler, where they are cooled to 450°C prior to going through a cyclone and a filter that retain most of the fly ash and the char particles, which are finally reinjected into the gasifier by pneumatic transport under nitrogen. The syngas goes successively through a water scrubber to remove halides and is desulfurized to be cleaned sufficiently to comply with the strict tolerance limits of fuel cells.

Texaco (now GE): The gasifier is a pressure vessel with a refractory lining that operates at temperatures in the range 1250 to 1450°C and pressures of 3 MPa for power generation and up to 6 to 8 MPa for H_2 and chemical synthesis. Figure 2.3 illustrates key features of the gasifier. The feedstocks, oxygen, and steam are introduced through burners at the top of the gasifier. Solid feedstocks such as coal are pre-processed into a slurry by fine grinding and water addition. The slurry is pumped into the burner, and the water, which is added with the slurry, replaces most of the steam that should normally be injected into the system. Raw gas and molten ash produced during coal gasification flow out toward the bottom of the gasifier. Two alternatives are then available for the recovery of the ash and for cooling the raw gas. The raw gas either can be cooled and cleaned from the slag ash by water quenching, or it can be cooled in a radiant syngas cooler from 1400 to 700°C. The heat recovered in the second option is then used

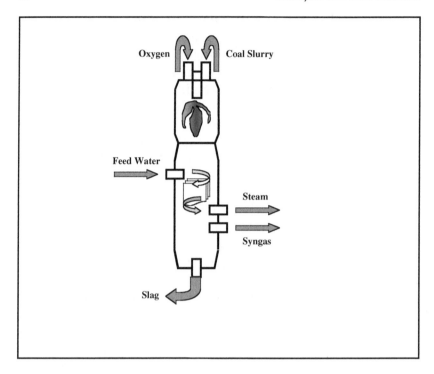

Figure 2.3 Key features of the Texaco gasifier.

to raise steam to be used in the process or for power generation. Molten slag flows down the heat-recovery steam generator and is quenched in a batch at the bottom of the cooler and finally removed through a lock hopper system. The quench alternative is the preferred option for coal feedstocks, as they could contain traces of salts (sodium and calcium) that could be corrosive for the syngas coolers at high temperatures. However, this alternative could be a disadvantage for power generation, as thermal efficiency is slightly lowered.

There are several existing projects based on Texaco technology, including an IGCC project, the Polk Power Station managed by Tampa Electric Corp. During the first three commercial years of operation, ten different coals or coal blends were tested to determine the cheaper feedstock to process while respecting new environmental regulations. The slag removal system of the Polk Power Station was designed for processing coals with a maximum of 12% (Wt dry basis) ash content. The operating temperature of the gasifiers has to be high enough for the coal mineral matter to melt and flow freely down the bottom of the gasifier. Texaco has fixed the minimum heating value of the coals at 30 MJ/kg to produce enough syngas to fully load the combustion turbine. It would be necessary to increase the oxygen supply size as well as the slurry delivery system capacity to be able to run the plant with a lower heating value coal. The plant is designed to accommodate coals with sulfur contents of up to 3.5% (Wt dry basis). Expensive modifications of the acid gas removal system were required for higher sulfur content coals than the first base coal (Pittsburgh No 8-1 with a sulfur content of around 2.5%). Following major problems, the company decided to switch to coal blends with a lower sulfur content. The limit in chlorine concentration in the coals was fixed at 0.15% (dry ash). A higher concentration of chlorine in coals would damage the system. Other coal properties have an influence on the technical and economic aspects of the Texaco-based IGCC operation and necessitate coal testing in the device prior to selecting them for the Polk Power Station.

The Texaco technology is also used for chemical plants. Five chemical plants were built after 1993 in the U.S. Eastman Chemicals (Kingsport, TN) owns two Texaco quench gasifiers that operate at 70 bar and 1400°C for the production of acetic acid and acetic anhydride. Although the facility is configured for the purpose of making acetyl chemicals, the company claims that gasification and cleanup plants are completely compatible with an electric power option, and in fact an electric power option of 523 MWe is reported to be under development at Kingsport. This is in line with the new projects of cogeneration of chemicals and electricity sponsored by the U.S. DOE under the Vision 21 program. Another U.S. company, Waste Management & Processors, Inc., is presently conducting a technoeconomic feasibility study in partnership with Texaco, Sasol, and Nexant for the development of one of the three demonstration Early Entrance Coproduction Plants (EECPs) under the Vision 21 program. The objective is the commercialization of a coal gasification/liquefaction technology to produce ultra-clean Fischer-Tropsch transportation fuels with power, chemicals, or steam as co-products. The proposed plant location is at the Gilberton Power Plant cogeneration facility in Gilberton, Pennsylvania. It involves the gasification of local waste coals, mainly high-ash-content anthracite wastes derived from an on-site coal cleaning operation that contains coal fines, coal dust, and dirt. Another demonstration EECP project is being developed by Texaco in collaboration with Rentech (Fischer-Tropsch Technology), Brown and Root Services, Praxair, and GE Power Systems for the production of electricity and chemicals from coal or petroleum coke. The project involves technical and economic studies of several process options, including syngas composition, Fischer-Tropsch product upgrading, wastewater treatment, catalyst/wax separation, acid gas removal, tail gas utilization, and site selection.

There is also a plan for the construction by coal power of a 430 MW IGCC plant based on the Texaco technology near the Hatfield colliery in the North of England. The IGCC project with CO_2 removal and production of H_2 is being studied by Jabobs Consultancy in cooperation with GE. The IGCC power plant is configured to be capable of removing 75% of the feed carbon as CO_2 prior to combustion in the gas turbine. By performing a 'sour shift' of the syngas, most of the carbon monoxide should be converted into carbon dioxide and an equal volume of hydrogen. If carbon dioxide removal is performed then the fuel for the combustion turbine will consist mainly of H_2.

BBP: Babcock Borsig Power (BBP) technology, also known as the Noell entrained flow technology, was first developed in 1975 in the former East Germany for the gasification of lignite in a 3 MW pilot plant. A full-scale (130 MW) gasifier was built in the 1980s to produce syngas and town gas. The technology was known as the GSP process before being acquired by Noell in 1991. The process can be a dry-fed or a liquid-fed oxygen-blown slagging gasifier. If solid fuel is to be gasified, it is first pulverized, then pneumatically conveyed to the feeding system, and dry fed together with oxygen and steam through a burner located at the top of the gasifier. Depending on the fuel ash content, the gasification chamber can be covered by either a cooling screen or a cooling wall. Both the refractory and the solid slag provide thermal insulation and maintain the tube surface temperature below 230°C. To allow the solidified slag to regenerate continuously, only fuels with an ash content of more than 1 wt%, such as coals, can be processed in the gasifier lined with a cooling screen. Heat removed by the cooled tube wall represents 2 to 3% of the total heat produced during gasification and is used to generate low-pressure steam. Syngas saturated with water is further cooled to 150 to 200°C and recycled to the quench sprays within the gasifier. The bottom part of the gasifier consists of a quench bath that cools and solidifies the slag, which is then removed in a granular form. The only Noell gasifiers in commercial operation are at Schwarze Pumpe (Germany) and the new BASF Seal Sands located in Middlesbrough in the UK. The BBP Research and Development center based at Freiberg (Germany) comprises two facilities with capacities of 5 and 10 MW. The smaller one was originally designed in 1979 for the gasification of both solid and pulverized solid materials. The pilot plant is presently used by Dow Chemical for the development of the gasification of chlorinated wastes. The second one was also designed for the gasification of pulverized materials (coal, waste), liquids, and slurries (waste oil, sludge, paint waste) and built in 1997. A wide range of coals from anthracites to brown coals have been gasified in the two pilot plants since the 1980s. BBP claims that it is capable of providing appropriate test conditions to optimize feedstock preparation prior to gasification as well as to determine the optimum gasification conditions for more than 80 different fuels, including 30 coals.

E-Gas: The E-Gas (formerly Destec) coal gasifier is a slurry-fed, pressurized, upflow, entrained two-stage slagging gasifier. Figure 2.4 illustrates key features. The dry coal concentrations in the slurry range from 50 to 70 wt%, depending on the inherent moisture and quality of the feed. Part of the coal slurry (80%) is injected with oxygen (95%) through two burners at the lower stage of the gasifier, where it is partially

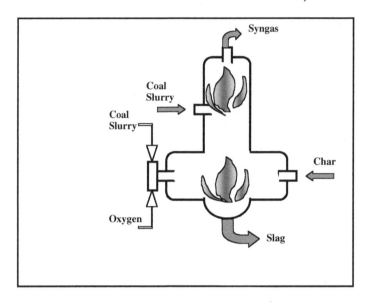

Figure 2.4   Key features of the E-Gas reactor.

combusted at a temperature of 1350 to 1400°C and a pressure of 3 MPa. Molten ash formed flows down the gasifier and is removed through a tap hole into a water quench. There is no lock hopper for ash removal, which has the advantage of reducing the overall height of the system. The fuel gas produced in the lower stage flows upwards into the upper stage, where it can react with the remaining 20% coal slurry. This two-stage process presents the advantage of producing a gas with a higher calorific value than the one produced in a one-stage process. The crude gas exiting the gasifier at a temperature of around 1050°C is cooled to 370°C in a firetube syngas cooler. This unit generates saturated high-pressure steam. The firetube syngas cooler is a boiler system, with the hot gas circulating on the boiler side, as opposed to a water syngas cooler, in which water circulates in tubes in a syngas tank. The firetube is reportedly considerably cheaper than the ones used in the Shell, Texaco, and Prenflo processes. After the cooling step the syngas is cleaned with filters to remove large ash and char particles that are pneumatically reinjected into the gasifier. The filter elements, made of metal for an acceptable resistance to corrosive syngas, are periodically back pulsed with high-pressure syngas to remove particulate cake formed on their surface. The particulate cake falls to the bottom of the vessel and is pneumatically transferred together with the high-pressure syngas to the first stage of the gasifier, where it is recycled. Finally, the particulate-free syngas proceeds to the low-temperature heat recovery system, where it is scrubbed with sour water condensed from the syngas to remove troublesome chlorides and trace elements that could cause corrosion within the piping and vessels as well as form undesirable products in the acid gas removal system.

After scrubbing and reheating, the syngas enters the COS hydrolysis unit, where the COS present in the syngas is converted to H_2S. The syngas is then cooled through a series of shell and tube exchangers to 35°C before entering the acid gas removal system. This cooling step also condenses water from the syngas. Most of the ammonia (NH_3) and some of the carbon dioxide (CO_2) and H_2S present in the syngas are absorbed in the water as dissolved gases. Wabash River (West Terre Haute, IN) is the only E-Gas gasifier in operation. Prior to the repowering of the Wabash River plant to an IGCC, some tests of bituminous coals, including high-sulfur coals, were performed in a 2200 ton coal/day plant based in Plaquemine, LA in the early 1990s. The Wabash River power plant is designed to use a range of local coals with a maximum sulfur content of up to 5.9% (dry basis) and a higher heating value of 31.4 MJ/kg (moisture and ash free). It is presently operating on Illinois No. 6 coal. Alternative fuels (petcoke) have also been successfully tested at Wabash River, and future tests may include coal fines. Coal fines are believed to be a promising fuel in the locality of the Wabash River facility as it is produced by the existing operations of the adjacent mine. They are also available from surface reserves, where the fines have been landfilled in the past and are predicted to be 40 to 60% cheaper than the present coal delivered to the facility.

Prenflo: Coal is fed together with oxygen and steam through four burners located at the lower part of the gasifier. Syngas is produced at a temperature of 1600°C and is quenched at the gasifier outlet with recycled, cleaned gas to reduce its temperature to 800°C. Then the syngas flows up a central distributor pipe and down through evaporator stages before exiting the gasifier at a temperature of 380°C. The raw gas is then dedusted in two ceramic candle filters, and a part of it is recirculated into the syngas cooler. The syngas is finally washed in a Venturi scrubber. Slag formed during the gasification process is quenched in a water bath and is then removed through a lock hopper system. The only commercial-scale unit is based in Puertollano in Spain (capacity of 338 MWe). It is the largest unit worldwide based on solid fuels. The plant has been operating since 1996 and can process up to 2600 ton/day of coal/petcoke fuel mixed with limestone (2% weight) and produces 180,000 m³/day of raw gas. The annual production of slag (85% of the ash in weight) and fly ash (15% of the ash in weight) are 120,000 ton and 12,000 ton per year, respectively. The demonstration project has now attained commercial development with a gross efficiency of 47.2% (net efficiency of 42%).

FLUIDIZED-BED TECHNOLOGIES

HTW: The High Temperature Winkler (HTW) process was first developed by Rheinbraun in Germany to gasify lignites for the production of a reducing gas for iron ore. The gasifier consists of a refractory-lined pressure vessel equipped with a water jacket. Feedstocks are pressurized in a lock hopper, which is located below the coal storage bin and then pneumatically conveyed to a coal bin. The conveying gas is then filtered and recirculated. Coal in the receiving bin is then dropped via a gravity pipe into the fluidized bed, which is formed by particles of ash, semi-coke, and coal. The gasifier is fluidized from the bottom with either air or oxygen/steam, and the temperature of the bed is kept at around 800°C, below the fuel ash fusion temperature. An additional gasification agent is introduced at the freeboard to decompose, at higher temperature (900 to 950°C), undesirable byproducts formed during gasification. The operating pressure can vary from 1 to 3 MPa, depending on the use of the syngas. The raw syngas produced is passed through a cyclone to remove particulates and then cooled. Solids recovered in the cyclones are rein-jected into the gasifier, and dry ash is removed at the bottom via a discharge screw. The syngas cooling system has been the subject of study as to whether to use a water-cooled or a firetube syngas cooler. The main reason was that the existing water-cooled syngas cooler was facing fouling and corrosion problems. A conventional water scrubber system was origi-nally used for gas cleaning but due to blockages, fouling, corrosion, and also the high operating cost of the system, Rheinbraun decided to develop a hot gas filtration system. A hot gas ceramic candle unit formed of 450 candles was devel-oped and operated for 15,000 hours.

The HTW technology manufactured by Rheinbraun was successfully applied for the synthesis of chemicals (methanol) from lignites at Berrenrath, Germany, between 1986 and 1997. The plant was shut down at the end of 1997 as, at the time, the process was no longer considered to be economically viable. Another commercial plant has been operating in Fin-land since 1988, essentially with peat for the production of ammonia. A 140 ton coal/day pressurized HTW gasification plant was also commissioned and built at Wesseling, Ger-many, in 1989, to supplement research and development of the HTW technology for coal use and particularly to study its future application to an IGCC process for power generation. The plant was designed for a maximum thermal capacity of 36 MW and was operated for 3 years either as an air-blown or an oxygen-blown gasification plant with pressures up to 2.5 MPa. A wide range of coals was tested in the Wesseling plant, including brown coals and a high-volatile bituminous coal (Pittsburgh No. 8). The Wesseling plant provided the operational data required to design a potential 300 MW com-mercial IGCC power plant (KoBra), which was finally never built. However, there is presently a project to develop a 400 MW IGGC plant based on the HTW technology (two units) to replace 26 existing Lurgi moving beds at Vresova in the Czech Republic. The new HTW plant (80 ton/hour coal and pressures up to 3 MPa) should operate on Czech lignite and will benefit from years of research and development at the Wesseling and Berrenrath plants. In order to adapt the HTW technology to the Czech lignites and also to the pre-existing Vresova IGCC plant (coal grinding plant, air separation unit, wastewater treatment, and steam turbine), tests were performed by Rheinbraun in an HTW bench-scale gasification unit and compared to results obtained with other coals in the same bench-scale unit and in a demonstration plant.

IDGCC: The Integrated Drying Gasification Combined Cycle (IDGCC) technology was specifically developed for the gasification of high-moisture, low-rank coals by Herman Research Pty Limited in Morwell, Australia. The gasifier is a 5 MW air-blown pressurized fluidized-bed pilot plant that is fed with coal from an integrated drying process. The feed coal is pressurized in a lock hopper system and then fed into the dryer, where it is mixed with the hot gas leaving the gasifier. The heat in the gas is used to dry the coal, while the evaporation of water from the coal cools down the gas without the need of expensive heat exchangers. The gasifier operates at 900°C under 2.5 MPa air pressure. Chars and ash are collected at the bottom of the gasifier and from a ceramic filter and burnt in a separate boiler. The final ash product is similar to that from a conventional low-rank boiler. A wide range of low-rank coals could be processed in the IDGCC, with only small changes in the operating conditions. Coals containing high levels of sulfur can be processed with sorbents, such as limestone or dolomite, directly injected into the bed. This would obviate the need for additional cooling of the gas to 40°C for sulfur removal from the very high-moisture syngas. The extra cooling would have led to a very large energy loss from water condensation and reduced mass energy for the gas turbine. It is expected that the IDGCC could handle coals with lower moisture content and higher ash content. As the IDGCC plant is based on a fluidized-bed gasification technology, it is then not recommended, as in most of the fluidized-bed technologies, for coals with relatively low reactivities and coals with low ash melting points. When looking at environmental considerations and particularly at the concept of CO_2 removal and H_2 production, the IDGCC, which produces a very moist syngas, can provide the water for the shift reaction without robbing or much reduced robbing of the steam cycle and may have potential for future development. It was reported that the IDGCC process is more efficient and as a consequence more environmentally friendly (lower CO_2 emission) than conventional processes, and would be just slightly less efficient than an Australian black coal IGCC process.

KRW: Figure 2.5 illustrates key features of this gasifier. Coal and limestone, crushed to below 1/4," are transferred from feed storage to the KRW fluidized-bed gasifier via a lock hopper system. Gasification takes place by mixing steam and

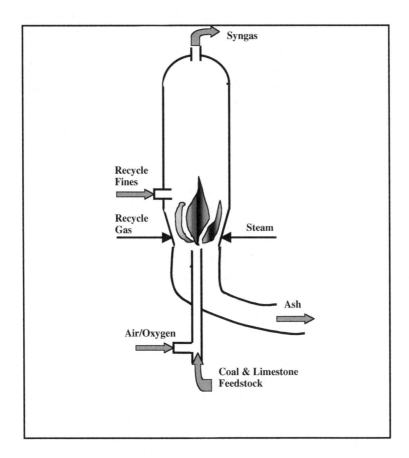

Figure 2.5 The KRW gasifier.

air (or oxygen) with the coal at a high temperature. The fuel and oxidant enter the bottom of the gasifier through concentric high-velocity jets, which ensure thorough mixing of the fuel and oxidant and of the bed of char and limestone that collects in the gasifier. After entering the gasifier, the coal immediately releases its volatile matter, which burns rapidly, supplying the endothermic heat of reaction for gasification. The combusted volatiles form a series of large bubbles that rise up the center of the gasifier, causing the char and sorbent in the bed to move down the sides of the reactor and back into the central jet. The recycling of solids cools the jet and efficiently transfers heat to the bed material. Steam, which enters with the oxidant and through a multiplicity of jets in the conical section of the reactor, reacts with the char in the bed, converting it to fuel gas. At the same time, the limestone sorbent, which has been calcined to CaO, reacts with H_2S released from the coal during gasification, forming CaS. As the char reacts, the particles become enriched in ash. Repeated recycling of the ash-rich particles through the hot flame of the jet melts the low-melting components of the ash, causing the ash particles to stick together. These particles cool when they return to the bed, and this agglomeration permits the efficient conversion of even small particles of coal in the feed. The velocity of gases in the reactor is selected to maintain most of the particles in the bed. The smaller particles that are carried out of the gasifier are recaptured in a high-efficiency cyclone and returned to the conical section of the gasifier, where they again pass through the jet flame. Eventually, most of the smaller particles agglomerate as they become richer in ash and gravitate to the bottom of the gasifier. Since the ash and spent sorbent particles are substantially denser than the coal feed, they settle to the bottom of the gasifier, where they are cooled by a counter-flowing stream of recycled gas. This both cools and classifies the material, sending lighter particles containing char back up into the gasifier jet. The char, ash, and spent sorbent from the bottom of the gasifier flow to the fluid-bed sulfator, where both char and calcium sulfide are oxidized. The CaS forms $CaSO_4$, which is chemically inert and can be disposed of in a landfill. Most of the spent sorbent from the gasifier contains unreacted CaO. Sulfur released from burning residual char in the sulfator is also converted to $CaSO_4$.

Pinon Pine in Nevada is the only large-scale coal-based IGCC plant (100 MWe) that is using the KRW technology, and it is also the only one that was designed with a 100% hot gas cleanup. The demonstration plant, owned by Sierra Pacific Resources and sponsored by the U.S. DOE, has had numerous problems. The gasifier had 18 start-ups, and all of them failed due to equipment design. Successes in the project included operation of the combined cycle portion of the plant at 98% availability, efficient removal by the hot gas filter of particulates from the syngas and production of a good quality syngas for only 30 hours since the first syngas was produced in 1998. Sierra Pacific Resources, which owns the Pinon Pine power plant, was going to be sold to WPS Power Development, but the sale has been suspended by the state of Nevada, which placed a moratorium on the sale of power plants in the state.

ABGC: The Air-Blown Gasification Cycle (ABGC) is a hybrid system that was developed at pilot scale (0.5 ton/hour coal capacity) by the former Coal Technology Development Division of British Coal. The gasifier is based on a spouted-bed design and is operated at pressures up to 2.5 MPa and a temperature between 900 and 1000°C. Coal fed in the gasifier produces a gas with a low calorific value of around 3.6 MJ/m^3. Sorbents such as limestone are also injected into the gasifier to retain up to 95% of the sulfur originally present in coal. Syngas is first cleaned in a cyclone, then cooled to around 400°C and cleaned by a ceramic filter, to be finally burned and expanded through a gas turbine. Only 70 to 80% of the fuel is gasified, and partially gasified char and other solid residues (fly ash and sulphided sorbent residues) produced in the gasifier are then transferred to an atmospheric pressure circulating fluidized-bed combustor (CFBC) operating at a temperature of about 1000°C. Heat generated by the combustion of the char supplies a steam cycle used to drive a steam turbine to supplement the electricity generation. The ABGC process is forecast to have an efficiency of about 46 to 48%. The ABGC technology was later purchased by Mitsui Babcock

Energy Limited (MBEL), which produced in collaboration with GEC Alsthom and Scottish Power PLC a design of a demonstration plant while being supported by the European Commission under the THERMIE program. A wide range of UK coals and international steam coals were studied for use in the ABGC. A laboratory at Imperial College of Science Technology and Medicine in London studied the impact of several coal characteristics on the gasification reactivity of some international traded coals in bench-scale reactors that could mimic the behavior of single coal particles in the ABGC. Coal characteristics studied included coal maceral composition and coal mineral matter composition.

BHEL: A 168 ton coal/day capacity air-blown pressurized fluidized-bed gasifier IGCC pilot plant (6.2 MWe) was built at Hyderabad, India, following previous gasification tests in an 18 ton coal/day capacity IGCC fluidized-bed gasifier pilot plant and in a 150 ton coal/day moving bed IGCC pilot plant. The plant consists of a refractory lined reactor with a 1.4 m inside diameter in the bed, expanding to a 2 m inside diameter at the upper section of the gasifier. Crushed coal (6 mm size or below) is injected into the system via a lock hopper and a rotary coal feeder and then pneumatically transported into the gasifier with a portion of the air used by the plant. The dry granular ash produced during gasification is withdrawn from the bottom of the gasifier through a water-cooled screw extractor and is discharged periodically through an ash lock system. Three refractory cyclones operating in series are used for primary gas cleaning. Fines collected in the first two cyclones can be recycled in the gasifier but there is also the possibility to collect the cyclone fines, without recycling, through a lock hopper. The gasifier operates at a temperature of 1000°C and pressure of 1.3 MPa to generate a coal gas with a net calorific value of 9.8 MJ/kg. The 168 ton coal/day demonstration plant was commissioned in 1996 and has since undergone a series of tests in standalone and in IGCC mode, operating for a total of 1200 hours until the year 2000. The plant is designed for the gasification of Indian coals with a high ash content of up to 42%.

Transport Reactor: The Kellogg Transport Gasifier is a circulating-bed reactor concept that uses finely pulverized coal and limestone. The gasifier is currently in development, which may lead to a commercial design. It is expected that the small particle size of the coal and limestone will result in a high level of sulfur capture. Additionally, the small particle size will increase the throughput compared to a KRW gasifier, thereby potentially reducing the required number of gasifier trains (or the gasifier size) and the cost. The Transport Gasifier is conceptually envisioned as consisting of a mixing zone, a riser, cyclones, a standpipe, and a non-mechanical valve. Oxidant and steam are introduced at the bottom of the gasifier in the mixing zone. Coal and limestone are introduced in the upper section of the mixing zone. The top section of the gasifier discharges into the disengager or primary cyclone. The cyclone is connected to the standpipe, which discharges the solids at

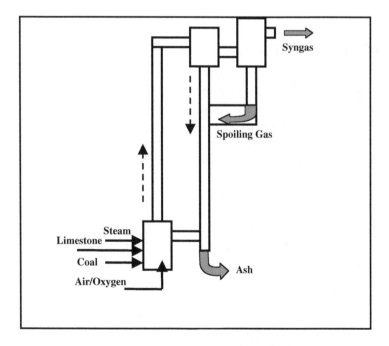

Figure 2.6 The Transport reactor process.

the bottom through a non-mechanical valve into the transport gasifier mixing zone at the bottom of the riser. The gasifier system operates by circulating the entrained solids up through the gasifier riser, through the cyclone, and down through the standpipe. The solids reenter the gasifier mixing zone through the non-mechanical valve. The steam and oxidant jets provide the motive force to maintain the bed in circulation and oxidize the char as it enters the gasifier mixing zone. The hot gases react with coal/char in the mixing zone and riser to produce gasification products. The gas and entrained solids leaving the primary cyclone pass through the secondary cyclone to provide final de-entrainment of the solids from the gas. The solids separated in the secondary cyclone fall through the dipleg into the standpipe. A solids purge stream is withdrawn from the standpipe for solids inventory maintenance. The gas leaving the secondary cyclone passes through a gas cooler, which reduces the gas temperature from about 1900°F to 1100°F.

MOVING-BED TECHNOLOGIES

BGL: The British Gas/Lurgi (BGL) coal gasifier is a dry-fed, pressurized, fixed-bed, slagging gasifier. The reactor vessel is water cooled and refractory lined. Each gasifier is provided with a motor-driven coal distributor/mixer to stir and evenly distribute the incoming coal mixture. Oxygen and steam are introduced into the gasifier vessel through sidewall-mounted tuyeres (lances) at the elevation where combustion and slag formation occur. The coal mixture (coarse coal, fines, briquettes, and flux), which is introduced at the top of the gasifier via a lock hopper system, gradually descends through several process zones. Coal at the top of the bed is dried and devolatilized. The descending coal is transformed into char, and then passes into the gasification (reaction) zone. Below this zone, any remaining carbon is oxidized, and the ash content of the coal is liquified, forming slag. Slag is withdrawn from the slag pool by means of an opening in the hearth plate at the bottom of the gasifier vessel. The slag flows downward into a quench chamber and lock hopper in series. The pressure differential between the quench chamber and gasifier regulates the flow of slag between the two vessels. Product gas exits the gasifier at approximately 1050°F through an opening near the top of the gasifier vessel and passes into a water quench vessel and a boiler feed water (BFW) preheater designed to lower the temperature to approximately 300°F. Entrained solids and soluble compounds mixed with the exiting liquid are sent to a gas-liquor separation unit. Soluble hydrocarbons, such as tars, oils, and naphtha, are recovered from the aqueous liquor and recycled to the top of the gasifier or reinjected at the tuyeres.

Lurgi: The Lurgi dry ash gasifier is a pressurized, dry ash, moving-bed gasifier. Sized coal enters the top of the gasifier through a lock hopper and moves down through the bed. Steam and oxygen enter at the bottom and react with the coal as the gases move up the bed. Ash is removed at the bottom of the gasifier by a rotating grate and lock hopper. The countercurrent operation results in a temperature drop in the reactor. Temperatures in the combustion zone near the bottom of the gasifier are in the range of 2000°F, whereas gas temperatures in the drying and devolatization zone near the top are approximately 500 to 1000°F. The raw gas is quenched with recycled water to condense tar. A water jacket cools the gasifier vessel and generates part of the steam to the gasifier. Sufficient steam is injected to the bottom of the gasifier to keep the temperature below the melting temperature of ash.

BHEL: The gasification media, a mixture of air and steam, is fed through a grate, which also enables ash removal. A gas cooler is used to recover part of the sensible heat of the gas produced and superheat steam for the gasifier. Further gas cooling as well as tar condensation are done by water quenching. Particulates are removed with a Venturi scrubber. A pilot plant has been operated for more than 5500 hours (1100 hours as IGCC), with two types of coals having high ash contents: Singareni coal with an ash content of 27 to 35% and North Karanpura coal with an ash content of 40%. The North Karanpura coal was also tested in the Lurgi pilot-scale plant at the Indian Institute of Chemical Engineering (IICT) under the same gasification conditions. It resulted in a better performance of the BHEL gasifier (calorific value and cold gas efficiency), due mainly to the larger scale of the gasifier. However, the availability of the plant was affected by the poor performance of the raw gas cooler due to tar deposition and choking. A direct contact quench was subsequently designed to replace the gas cooler and overcome that problem. The performance of the moving-bed gasifier was also compared to that of a pressurized fluidized-bed gasifier later developed by BHEL at the Trichy unit in Hyderabad in India. Moving-bed gasifiers produce tar-laden gas, which make the recovery of the sensible heat of the raw gas difficult. They also need coals with a certain particle size (5 to 30 mm). They produce large effluents containing tars and phenolic acids, requiring elaborate effluent treatment. For these reasons, BHEL decided to develop the fluidized-bed technology for the processing of Indian coals. A 6.2 MWe IGCC plant was developed by BHEL at the Trichy unit in 1988, as part of a research program for the development of gasification of Indian coals for the production of electricity. The gasification process was based on a moving-bed technology developed in-house, after experience on a Lurgi dry ash bed gasifier (pilot-scale 24 ton/day) was gained at the Indian Institute of Chemical Engineering at Hyderabad and at CFRI at Dhanbad. The gasifier is a 2.7 m diameter, 14 m high jacketed moving-bed gasifier with a coal throughput of 150 tpd. Crushed coal of 5 to 40 mm size with an ash content of about 35% is the design feedstock for the gasifier, which is operating at 1 MPa pressure.

TECHNOLOGY SUPPLIERS

Different technology suppliers worldwide have developed gasifiers. The choice of the type of the gasifier is purely a factor of the coal/fuel characteristics. Various technology suppliers for the gasification process are listed in Table 2.2.

SYNGAS CHARACTERISTICS

Composition of the syngas depends on the fuel as well as on the gasification process. The typical characteristics of the syngas

TABLE 2.2   Gasification Technology Suppliers

Technology Supplier	Coal Feed Type	Oxidant	Gasifier Type
Texaco, U.S.	Water slurry	O_2	Entrained flow
Shell, U.S.	N_2 carrier/dry	O_2	Entrained flow
KRW, U.S.	Dry	Air	Fluidized bed
Lurgi, Germany	Dry	Air	Fluidized bed
British Gas/Lurgi	Dry	O_2	Moving bed
Prenflo, U.S./Krupp Uhde, Germany; Deutsche-Babcock, Germany	Dry	O_2	Entrained flow
Destec Energy, U.S.	Water slurry	O_2	Entrained flow
IGT U-Gas, U.S./Carbona, Finland/ IBIL, India	Dry	Air	Fluidized bed
Rheinbraun HTW, Germany/RWE Energie, Germany	Dry	Air	Fluidized bed
MHI, Japan/IGC, Japan	Dry	Air/O_2	Entrained flow
ABB-CE, U.S.	Dry	O_2	Entrained flow
VEW/Steinmuller, Germany	Dry	O_2	Entrained flow
Hitachi, Japan	Dry	O_2	Entrained flow
Noell/GSP	Dry	O_2	Entrained flow
Ahlstrom, Sweden	Dry	Air	Fluidized bed

as generated from different fuels at some of the IGCC projects are summarized in Table 2.3.

GAS CLEANUP SYSTEMS

The typical steps for a gas cleanup system aim at particulate removal, sulfur removal, and NO_x removal. This is achieved as follows:

- Particulate Removal: Combination of cyclone filters and ceramic candle filters
- SO_x and NO_x removal: Combination of steam/water washing and removing the sulfur compounds for recovery of sulfur as a salable product

Hot gas cleanup technology is currently under demonstration phase, but various demonstrations have not been successful so far. Wet scrubbing technology, though with a

TABLE 2.3　Typical Syngas Compositions

	Project						
Fuel	PSI Wabash	Tampa Polk	El Dorado	Shell Pernis	Sierra Pacific	IBIL	Schwarze Pumpe
	Coal	Coal	Pet coke/waste oil	Vacuum residue	Coal	Lignite	*
H	24.8	27.0	35.4	34.4	14.5	12.7	61.9
CO	39.5	35.6	45.0	35.1	23.5	15.3	26.2
CH_4	1.5	0.1	0.0	0.3	1.3	3.4	6.9
CO_2	9.3	12.6	17.1	30.0	5.6	11.1	2.8
N_2+Air	2.3	6.8	2.1	0.2	49.3	46.0	1.8
H_2O	22.7	18.7	0.4	—	5.7	11.5	—
LHV, kJ/M^3	8350	7960	9535	8235	5000	4530	12,500
T_{fuel}, °C	300	371	121	98	538	549	38
Oxidant	O_2	O_2	O_2	O_2	Air	Air	O_2

* Lignite/oil slurry with waste plastic and waste oil

lower efficiency, still remains the preferred option for gas cleanup systems in IGCC.

TECHNOLOGY SUPPLIERS FOR PARTICULATE REMOVAL

Table 2.4 summarizes key manufacturers of particulate removal technologies.

Sulfur Removal

Sulfur from the hot fuel gas is captured by reducing it to H_2S, COS, CS_2, and so on. The current sulfur removal systems employ zinc-based regenerative sorbents (zinc ferrite, zinc titanate, and so on) Such zinc-based sorbents have been demonstrated at temperatures up to 650°C. Sulfur is also removed by the addition of limestone in the gasifier. This is commonly

TABLE 2.4   Manufacturers of Gas Cleanup Systems

Manufacturer	Gas Temp. (Max.)	Particle Collection Efficiency	Remarks
Westinghouse Ceramic Candle Filter	1000°C	99.99% for 0.1 mm size	Hanging type candles
LLB Lurgi Lentjes Babcock Ceramic Candles Filter	1000°C	99.99%	Supported both sides
Pall Process Filtration Ceramic Candle Filter	1000°C (max.)	99.99%	Supported both sides; clay bonded silicon carbide filter
Schumacher Ceramic Candle Filter	1000°C	99.9%	Hanging type candles; clay bonded silicon carbide filter
Mott Metal Candle Filter	950°C	99.99%	Hanging type candles; sintered hastelloy

adopted in air-blown fluidized-bed gasifiers. In the case of air-blown gasifiers, sulfur is captured in the gasifier bed itself (above 90%) because of the addition of limestone. The sulfur captured in the bed is removed with ash.

The Power Block

The Power Block in the IGCC plant is essentially a gas turbine unit that operates on syngas. This gas turbine unit is basically the same as that used for natural gas with certain modifications. The areas that are modified and also that need to be critically evaluated for use with syngas are:

- Modification of fuel supply system
- Modification in the burners; special burners are required when using syngas because of its higher flame propagation velocity
- Checking for surge conditions and suitability of gas turbine units because of excess flow in case of syngas on account of it being a lean gas

The gas turbine/combined cycle (GTCC) technology has been proven for use with natural gas as well as with syngas.

COMPARISONS BETWEEN TECHNOLOGIES

Tables 2.5 and 2.6 provide a database on the key features of the technologies considered in this study along with the producer gas heating values and compositions. A considerable amount of data is missing, which is indicative of the need for comprehensive performance information to be organized into design practice literature to guide developers. These tables do, however, form the starting basis for the analysis presented in the sections that follow.

SYNGAS APPLICATIONS AND TECHNOLOGY SELECTION CRITERIA

Syngas composition varies based on many factors, including reactor type, feedstock, and processing conditions (e.g., temperature, pressure, type reactant, etc.). In turn, specific end-use

TABLE 2.5 Summary of Key Characteristics of Gasification Technologies

Coal Gasification Technologies

Entrained Flow Geometry

Developer/Process	Feedstock	Syngas Gas HV	Application	Feed Rate, tpd	Primary Reaction Press., psi	Primary Reaction Temp., °C	Secondary Reaction Press., psi	Secondary Reaction Temp., °C	Reactant	Gas Exit Temp., °C
Hitachi Coal Energy Application for Gas, Liquid, and Electricity (EAGLE)	Coal	MHV	IGCC, Integrated Coal Gasification Fuel Cell (IGFC) combined cycle	150	360				Oxygen or air	450
Shell Coal Gasification Process (SCGP)	Coal (bitumous, hi moist./hi ash lignites)	LHV	IGCC, ammonia, urea, H₂, methanol; 400 MW under development	220, 365 (hi moist.), 2000	290–580	1500			Steam/O₂	300
Mitsubishi Heavy Industries (MHI)	Coal (high ash m.p. Australian)	LHV	IGCC: units developed include 27, 125, 250 MW	200, 1500	Amb.				Oxygen	350–450
Texaco	Coal	MHV	IGCC (430 MW and 523 MW plants), H₂, chemical synthesis	2300	435 (Power) 870–1160 (H₂), 1,015 (acetic acid, acetic anhydride)	1200–1450			Steam/O₂	700

TABLE 2.5 (continued)   Summary of Key Characteristics of Gasification Technologies

Developer/ Process	Feedstock	Syngas Gas HV	Application	Feed Rate, tpd	Primary Reaction		Secondary Reaction			Gas Exit Temp., °C
					Press., psi	Temp., °C	Press., psi	Temp., °C	Reactant	
Babcock Borsig Power (Noell)	Coal (anthracites and brown), waste oil, sludge	LHV	IGCC (experience gained on 5, 10, and 30 MW plants)						Steam/O$_2$	150-200
E-Gas (Destec) (see Kentucky IGCC project.)	Coal (bituminous, hi S, IL. No. 6, Petcoke)	LHV	Production of steam, fuels/chemicals, and electricity, IGCC (96, 296 MW plants)	2200	435	1350–1400				370
Prenflo	Coal-Petcoke (50/50)	LHV	IGCC (largest plant 338 MW)	2600	363				Oxygen	380
Fluidized Bed Geometry										
Integrated Drying Gasification Combined Cycle (IDGCC)	Coal (hi moisture, low rank)	LHV	IGCC		363	900			Air	40
Air-Blown Gasification Cycle (ABGC)	Coal	LHV	IGCC		363	900–1000	atm.	1000		400

BHEL (Indian Institute of Tech.)	Coal (hi ash — 42%)	LHV	IGCC (6.2 MW pilot plant basis for development)	18, 150, 168	188	1000		
High Temperature Winkler (HTW)	Coal (lignites, hi volatile bituminous), coal/coke mix, peat	MHV	Methanol, ammonia, IGCC (built 36 and 300 MW, designed 400 MW)	140	145–435	800		Steam/O_2, air
Kellog Rust Westinghouse (KRW)	Coal	LHV	IGCC (100 MW plant)		290	900	600	Air/Steam
Transport Reactor Gasifier	Coal (sub-bituminous), KY and IL No. 6, coke breeze	LHV	IGCC	26.8–64		870–1000	217	O_2, Air
Moving Bed								
BHEL pilot plant	Low-rank coals	LHV	IGCC (plant sizes developed between 6 and 150+ MW)	24.0, 150	145			Air/Steam
Lurgi dry ash process	Coal (bituminous, lignites)	LHV	Fuels, chemicals, ammonia (several pants 100 MW plus 361 MW)	18,000	363–406	980–1040	300–500	Steam/O_2
Schwarze Pumpe complex (Germany)	Coal (bituminous), MSW	MHV	Electricity (85 MW), methanol					

TABLE 2.5 (continued) Summary of Key Characteristics of Gasification Technologies

Developer/ Process	Feedstock	Syngas Gas HV	Application	Feed Rate, tpd	Primary Reaction		Secondary Reaction		Reactant	Gas Exit Temp., °C
					Press., psi	Temp., °C	Press., psi	Temp., °C		
British Gas/Lurgi (BGL)	Coal, sewage, sludge, MSW	LHV	IGCC			2000			CO_2, O_2, Steam	450–500
Biomass Gasification Technologies										
Updraft Fixed Bed										
Wellman Process Engr.	Wood	LHV	Engine	10	NA	600–1000			Air/Steam	100
Volund (Ansaldo)	Straw	LHV	Heat	13	atm.	NA			Air/Steam	250
Union Carbide Purox Process	MSW	MHV	IGCC	200		750–1100			Oxygen	180–300
Sofresid-Caliqua	MSW	LHV	Steam for district heating and electricity	215	14.5	1300–1400			Hot air	
Bubbling Fluid Bed										
Gas Techn. Inst.	Woody biomass	MHV	Fuel gas, syngas	3.6–12	479	816			Oxygen/ Steam	816
MTCI	Pulp, paper mill sludge	MHV	Steam	7	15	790–815			Steam	
Alternate Gas (Citicorp Ind. Credit)	Wood chips	LHV	Lime kiln, boiler and drier fuel	200	14.7	649–815			Hot Air	745–801

Energy Products of Idaho (formerly JWP Energy Products)	Wood chips	LHV	Steam for power production	110	14.7	650			Air	621
ASCAB/Stein Industrie	Wood chips	LHV	Methanol production, electricity (process has been abandoned)	50	220.5	716	220.5	1300	Steam/O_2	
Tampella Power Inc.	Biomass, coal	LHV	Fuel for gas turbines, boiler fuel	40	290–334	850–950			Air	300–350
BECON Iowa State (Biomass Energy Conservation Facility)	Shelled corn	LHV	IGCC	5	14.7	730			Air	
Circulating Fluid Bed										
BCL/FERCO	Wood	MHV	Fuel gas	26–200 (200 considered min. acceptable size)	15	600–1000		1300	Air/Steam	820
Thermal Process Studsvik (TPS) (Studsvik Eneritekmik AB)	Woody biomass	LHV	Fuel gas, duel fuel engine, gas turbine, boiler/furnace/kiln fuel	13–78 (upper limit commercial)	14.7	700–900	14.7	850–950	Air	
Lurgi Energy	Bark	LHV	Lime kiln firing	120	14.7	800			Air	600

TABLE 2.5 (continued) Summary of Key Characteristics of Gasification Technologies

Developer/ Process	Feedstock	Syngas Gas HV	Application	Feed Rate, tpd	Primary Reaction		Secondary Reaction			Gas Exit Temp., °C
					Press., psi	Temp., °C	Press., psi	Temp., °C	Reactant	
Aerimpianti (subsidiary of Ansaldo)	RDF	LHV	Cement kiln firing	48–110	7.25	850–900			Air	800–900
Foster Wheeler (formerly Ahlstrom)	Wood	LHV	Lime kiln firing, electricity production	16	14.7	905			Air	700
Sydkraft AB (in cooperation with Foster Wheeler)	Wood	LHV	IGCC — electricity and district heating (6 MWe and 9 MWt)		261	950–1000			Air	

TABLE 2.6 Calorific Value and Compositions of Syngas

Developer/Process	Heating Value		Syngas Composition (Wt%)									
	MJ/m³	Btu/ft³	H_2	CO	CO_2	CH_4	C_2^+	S as H_2S	H_2O	N_2	Other	CO/H_2
Entrained Flow Geometry												
Hitachi EAGLE	10.4	280										
Shell Coal Gasification Process (SCGP)	8.2	221	34.4	35.1	1–5	0.3			—	0.2		2.0–2.6
Mitsubishi Heavy Industries (MHI)	4.187	112										
Texaco	10–12	268–321										
Babcock Borsig Power (Noell)												
E-Gas (Destec)	10.3	277	34.4	45.3	15.8	1.9				1.9		
Prenflo	est. <10	est. <268										
Fluidized Bed Geometry												
Integrated Drying Gasification Combined Cycle (IDGCC)									Hi			
Air-Blown Gasification Cycle (ABGC)	3.6	96										
BHEL (Indian Institute of Tech.)												
High Temperature Winkler (HTW)												

TABLE 2.6 (continued) Calorific Value and Compositions of Syngas

Developer/Process	Heating Value		Syngas Composition (Wt%)									
	MJ/m³	Btu/ft³	H₂	CO	CO₂	CH₄	C₂⁺	S as H₂S	H₂O	N₂	Other	CO/H₂
Kellog Rust Westinghouse (KRW)												
Transport Reactor Gasifier	4	107										
Moving Bed												
BHEL pilot plant												
Lurgi dry ash process												
Schwarze Pumpe complex (Germany)												
British Gas/Lurgi (BGL)												
Biomass Gasification Technologies												
Updraft Fixed Bed												
Wellman Process Engr.	5.5	148	6.9	29.5	6.1		22.2			35		
Volund (Ansaldo)	2.6–5.0	70–134	4.4	11.6	14.7	4				64		
Union Carbide Purox Process	13.7	367	23.43	39.06	24.41	5.47	4.93	0.05			2.65	
Sofresid-Caliqua	Low		NA	NA	NA	NA	NA	NA	NA			

Bubbling Fluid Bed											
Gas Techn. Inst.	12.9	350	25.3	16	39.4	17.8	1.5	—	—	0	—
MTCI	16.2	438	43.3	9.22	28.1	4.73	9.03		5.57	48	0.08
Citicorp Ind. Credit	6.9	186	12.67	15.5	15.88	5.72	2.27		0	52	0.8
Energy Products of Idaho	5.6	150	5.8	17.5	15.8	4.65	2.58	0	0		
ASCAB/Stein Industrie	5.5	155	19.87	25.3	40	0	0			13	
Tampella Power Inc.	5	140	11.3	13.5	12.9	4.8			17.7	40	
BECON Iowa State	4.5	126	4.1	23.9	12.8	3.1				56	0.2
Circulating Fluid Bed											
BCL/FERCO	18.7	500	14.9	46.5	14.6	17.8	6.2	—	10–14	47–52	0.5–1.0
Thermal Process-Studsvik (TPS)	5.5	147	7–9	9–13	12–14	6–9	—	—			
Lurgi Energy	5.8	155	20.2	19.6	13.5	3.8				43	0.1
Aerimpianti	5	134	7–9	9–13	12–14	6–9			10–14	47–52	0.5–1.0
Foster Wheeler	7.5	201	15–16	21–22	10–11	5–6				46–47	
Sydkraft AB	5.8	121	9.5–12	16–19	14.4–17.5	5.8–7.5				48–52	

product applications depend on the intermediate products of the syngas and the limitations of technologies for syngas conversion to fuels. Unfortunately, direct comparison between biomass-based and coal-based technologies is not straightforward, and indeed even comparisons among coal-based gasification technologies are not a simple task because of the wide variation in coal feedstock compositions used by developers. Despite these shortcomings, comparisons are possible, especially when we recognize that biomass gasification technologies in general produce a lower quality syngas when compared with coal gasification under similar operating conditions.

Table 2.7 provides the criteria applied in matching various technologies to applications. Note that for completeness, Table 2.7 includes the desirable syngas characteristics for synthetic fuels.

Straight combustion of coal-based syngas fuels in boilers is a fully developed technology with likely few opportunities for expanded coal utilization in the long-term. There are few advantages to such systems from an environmental standpoint as process schemes simply move environmental controls upstream from the boiler. Steam boilers can tolerate some levels of contaminants, including chlorines, particulates, and sulfur.

Commercialization and R&D efforts largely focus on either co-firing biomass-derived syngas in coal-fired boilers, and on co-feeding biomass and coal options. Co-firing technologies under development focus on biomass gasification technologies, where the syngas simply plays the same role as natural gas in a co-fired coal-NG (natural gas) boiler. Co-feeding schemes are too new to assess their merits, but their primary advantage is fuel flexibility. These systems, which are largely conceptual at this point, have the advantage of fuel flexibility, primarily in favor of biomass as there are concerns for feedstock supplies. These systems are well suited for the 50 MW range simply because biomass gasification technologies tend to be either cost prohibitive or require too much feedstock handling above about 100 MW.

TABLE 2.7 Desirable Syngas Characteristics for Different Applications Based On Current Technology Limitations

| Product | Synthetic Fuels | | | Boiler | Turbine | Fuel Gas | | | |
| | FT Gasoline and Diesel | Methanol | Hydrogen | | | Fuel Cell | | | |
						PAFC	MCFC	SOFC	PEFC
H_2/CO	0.6^a	~2.0	High	Unimportant	Unimportant	H_2 is fuel[n] but CO is a poison >0.5%	Both H_2 and CO are fuels, with H_2 preferred[l]	H_2 and CO are fuels	H_2 is fuel but CO is a poison >10 ppm
CO_2	Low	Low[c]	Not important[b]	Not critical	Not critical	Diluent	Diluent	Diluent	Diluent
Hydro-carbons	Low[d]	Low[d]	Low[d]	High	High	CH_4 is diluent	CH_4 is diluent[m]	CH_4 is fuel[o]	CH_4 is diluent
N_2	Low	Low	Low	Note[e]	Note[e]				
H_2O	Low	Low	High[f]	Low	Note[g]				
Contami-nants	<1 ppm sulfur and low particu-lates	<1 ppm sulfur and low particu-lates	<1 ppm sulfur and low particu-lates	Note[k]	Low particu-lates and low metals	S as H_2S and COS is a poison >50 ppm	S as H_2S and COS is a poison >0.5 ppm	S as H_2S and COS is a poison >1.0 ppm	No studies to date
Heating value	Unimportant[h]	Unimportant[h]	Unimportant[h]	High[i]	High[i]	Unimpor-tant	Unimpor-tant	Unimpor-tant	Unimpor-tant

TABLE 2.7 (continued) Desirable Syngas Characteristics for Different Applications Based On Current Technology Limitations

| | Synthetic Fuels | | | Fuel Gas | | Fuel Cell | | | |
| | FT Gasoline | | | | | | | | |
Product	and Diesel	Methanol	Hydrogen	Boiler	Turbine	PAFC	MCFC	SOFC	PEFC
Pressure, bar	~20–30	~50 (liquid phase) and ~140 (vapor phase)	~28	Low	~400	Up to 125 psi	High	216–441 psi	High
Temperature, °C	200–300j; 300–400	100–200	100–200	250	500–600	100–200	650	1000	80

Footnotes to Table 2.7:

(a) Depends on catalyst type. For iron catalyst, value shown is acceptable; for cobalt catalyst, a value closer to 2.0 is recommended.

(b) Water-gas shift is needed to convert CO to H_2; CO_2 in syngas can be removed at same time as CO_2 by the water in the gas shift reaction.

(c) Some CO_2 can be tolerated if the H_2/CO ratio > 2.0 (as can occur with steam reforming of natural gas); if excess H_2 is available, the CO_2 will be converted to methanol.

(d) Methane and heavier hydrocarbons must be recycled for conversion to syngas and represent system inefficiency.

(e) N_2 lowers the heating value; however, the level is unimportant provided the syngas can be burned with a suitable flame.

(f) Water is required for the water-gas shift reaction.

(g) Capable of tolerating high water levels; steam is sometimes added to moderate combustion temperature for NO_x control purposes.

(h) As long as H_2/CO and impurities levels are met, heating value is not critical.

(i) Efficiency improves as heating value increases.

(j) Depends on the catalyst type; iron catalysts typically operate at higher temperatures than cobalt catalysts.

(k) Small amounts of contaminants can be tolerated.

(l) In reality, CO with H_2O shifts H_2 and CO_2, and CH_4 with H_2O reforms to H_2 and CO faster than reaction as a fuel at the electrode. CO is a poison for lower temperature fuel cells, but is used as a fuel in the high-temperature cells (e.g., SOFC, MCFC). CO may not actually react electrochemically within these cells. It is commonly understood that CO is consumed in the gas phase through the water-gas shift reaction as $CO + H_2O = CO_2 + H_2$. The H_2 formed in this reaction is subsequently consumed electrochemically.

(m) CH_4 is a fuel in the internal reforming stage of MCFC.

(n) H_2 is the optimal fuel for all types of fuel cells.

(o) CH_4 can be oxidized directly using a solid oxide fuel cell; however, high concentrations of CH_4 lead to severe coking problems. Only cells containing dilute concentrations of CH_4 can be oxidized directly in current SOFCs. In addition, the oxidation of CH_4, like that of CO, may not actually occur at active electrochemical sites within an SOFC. Rather, CH_4 is probably reformed within the cell through steam reforming.

Integrated Gasification Combined Cycle

The IGCC process is a two-stage combustion with cleanup between the stages. The first stage employs the gasifier, where partial oxidation of the solid/liquid fuel occurs by limiting the oxidant supply. The second stage utilizes the gas turbine combustor to complete the combustion, thus optimizing the gas turbine/combined cycle technology with various gasification systems. The syngas produced by the gasifiers, however, needs to be cleaned to remove the particulate, as well as wash away sulfur compounds and NO_x compounds before it is used in the gas turbine. It is the integration of the entire system components, which is extremely important in an IGCC plant. Various sub-systems of an IGCC plant are:

• The gasification plant
• The power block
• The gas cleanup system

Recent advances in the gas turbine technologies have presented great potential towards much higher gas turbine efficiencies. Increasing the firing temperatures and utilizing materials that withstand higher temperatures can increase the efficiency of a gas turbine. Continuous developments have been taking place in the newer materials of construction, thus enabling higher gas turbine performance. At present the efficiency of gas turbines is in the range of 45 to 50%, which is projected to go up to 60% with the development of H-technology by GE. The advances in gas turbines would improve the overall efficiency of IGCC plants. The World Bank has provided a comparative study showing the likely achievable improvements of IGCC power plant efficiency, which is summarized in Figure 2.7.

Lower Heat Rates and Increased Output. The heat rates of the plants based on IGCC technology are projected to be around 2100 kCal/kWh compared to the heat rates values of around 2500 kCal/kWh for the conventional PC fired plants.

Flexibility to accept a wide range of fuels. IGCC technology has been proven for a variety of fuels, particularly

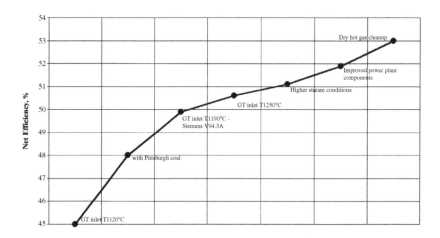

Figure 2.7   Projected improvements of IGCC power plant efficiency.

heavy oils, heavy oil residues, petcokes, and bituminous coals in different parts of the globe. In fact the same gasifiers can handle different types of fuels.

Environment Friendly Technology. IGCC is an environmentally benign technology. The emission levels in terms of NO_x, SO_x, and particulate from an IGCC plant have been demonstrated to be much lower when compared to the emission levels from a conventional PC fired steam plant.

The technology level for each individual system component of IGCC, i.e., gasification block, gas cleanup system, and power block have already been established and proven in practice at the commercial level. Integrating these individual technologies for the electricity generation is the concept of IGCC. To demonstrate IGCC technology at the commercial level, a number of projects have been in demonstration/operation stage. The fact that the IGCC technology has reached maturity stage can be seen from Tables 2.8 through 2.9, which give the status of various IGCC projects.

Table 2.10 provides some case studies of IGCC projects that are described on the following url address: http://europe. eu.int/comm/energy_transport/atlas/html/cccasestudies.html.

TABLE 2.8 Partial Listing of Coal-Based IGCC Projects Worldwide

Project	Capacity	Operation	Fuel	Remarks
Coolwater Plant, Barstow,Ca, U.S.	125 MW	Operated 1984–88		GE (Texaco) gasifier (1000 TPD)
Plaquemine Plant, Louisiana, U.S.	160 MW	In operation since April 1987		Dow (Destec) gasifier (2200 TPD)
Demkolec Buggenum Plant, Netherlands	253 MW	Started operation in 1993		Shell gasifier – Initial problems encountered in gas cleanup system. Now operating with good availability.
PSI Energy, Wabash River Plant, U.S.	262 MW	Commissioned November 1995		Destec gasifier, repowering plant
Tampa Electric Polk Power Plant, U.S.	260 MW	Commissioned September 1996		GE (Texaco) Gasifier
Sierra Pacific Pinon Pine Plant, U.S.	100 MW	Commissioned 1998		KRW Gasifier No longer operating
ELCOGAS, Puertollano, Spain	335 MW	Prenflow, Krupp Uhde		Prenflo gasifier commissioned in 1998
Schwarze Pumpe, Germany	40 MW	Noell KRC (7 fixed-bed gasifiers)		Commissioned on syngas September 1996 power/methanol

TABLE 2.9 Partial Listing of Refinery Residue Based IGCC Projects Worldwide

Project	Capacity	Gasifier	Fuel	Status
Luenen, STEAG	170 MW	Lurgi oil		Operated 1972–1977
Texaco, El Dorado, U.S.	40 MW	GE (Texaco)	Waste/pet coke	Commissioned September 1996, Cogen
ILVA, Taronto, Italy	500 MW	—	Mill recovery gases	Commissioned January 1997
Shell PER+, Pernis, Netherlands (IGCC retrofit)	127 MW	Shell SGHP process	Heavy residues	Commissioned on NG in June 1997 and on syngas in November 1997
ISAB, Sicily, Italy	520 MW	GE (Texaco)	Asphalt	Passed acceptance test in January 2001
Sarlux, Sardinia, Italy	551 MW	GE (Texaco)	Refinery residue	Commissioned early 2000 (3 GE MS9001E GTs)
API-Energia, Falconara, Italy	280 MW	GE (Texaco)	Refinery residue	Passed acceptance test in April 2001
Star, Delaware, U.S./Saudi Aramco-Texaco JV	240 MW	GE (Texaco)	Pet coke	Commissioned August 2000, Cogen (120 MWe + steam), Repowering, 2 X GE 6FA GTs
Negishi	361 MW	GE (Texaco)	Heavy oil	Commercial operation started in 2003

Table 2.10 Partial Listing of Planned IGCC Projects Worldwide

Project/Location	Capacity	Gasifier	Fuel	Status
Mesaba Energy/U.S.	532 MW	ConocoPhillip's E-Gas (previously owned by Destec)	Coal	Announced — commercial operation in 2010
Rafineria Gadansk/ Poland	1600 tonne/ day	Shell	Asphalt	Start-up scheduled for 2006 — will produce power and H_2
ENI Sannazzaro/ Italy	50 tonne/hr	Shell	Tar	Under construction in 2004 — will produce power and H_2
Orlando Utility Commission's Stanton Energy Center/U.S.	285 MW	Transport	Coal	Announced — commercial operation in 2010
Bio Electrica/Italy	12 MW	Lurgi	Biomass	Commercial operation 2005
AEP/U.S.	Up to 1,000 MW	To be decided	Coal	Planned to be operating by 2010
Versova/Check Republic	400 MW	HTW	Lignite	Commissioning scheduled for 2005

Operational Feedback

Typical problems that have been encountered in various projects relate to the following areas:

- Gas Turbine Combustors: GT combustor design has been altered to handle low BTU gas with high mass flow due to problems encountered in gas turbines.
- Hot Gas Cleanup System: Breakage of ceramic candle filters and stress corrosion cracking in heat exchangers has also been reported.

Investment Costs

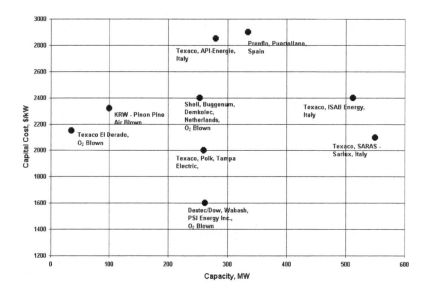

Figure 2.8 Capital costs reported for different demonstration/commercial projects.

The costs for the IGCC-based plants are variable, depending on economy of scale, local labor costs, and applicable engineering standards. Further, gasification costs usually are estimated in combination with the downstream processing equipment necessary for delivery of a syngas suitable for conversion to the designed end product. Accordingly, gasification investment costs are best addressed on a project-specific basis. The typical project costs as reported for different demonstration/commercial projects are summarized in Figure 2.8.

GUIDE TO COMMERCIAL EXPERIENCE

Table 2.11 provides a summary of the commercial and semi-commercial experiences for gasification systems.

Table 2.12 provides a survey of the coal gasification technologies currently under development.

TABLE 2.11 Gasification Technology Demonstration/Pilot Plant Scale

Technology Developer/Location	Capacity	Status	Fuel	Comments
SCGP-1, Shell Oil, Deer Park Complex, Texas, U.S.	250 TPD demonstration unit	Operated between 1987–1991	Coal/ lignite/ pet coke	Shell gasifier - 80% coal to clean gas efficiency, 99% sulfur removal achieved. Gas used for synthesis.
Rheinische Braunkohlenwerke, Berrenrath, Germany	720 TPD		Dry lignite	HTW gasifier; gas used for methanol production
American Natural Gas Co., Beaulah, North Dakota, U.S.	1000 TPD	In operation since 1984	Lignite	14 Lurgi dry ash gasifiers of 1000 TPD each for syngas production
British Gas Lurgi Westfield, Scotland	600 TPD/30 MWe	Commissioned in 1984		Demonstration unit; pressurised dry feed moving bed slagging BGL Gasifier
IGT U-Gas Shanghai, China	800 TPD (8 trains)	Commissioned in December 1994	Coal	1st IGT U-Gas commercial plant, industrial fuel gas
Krupp-Koppers, Saarbrucken, Germany	48 TPD		Coal	
IGT RENUGAS Maui, Hawaii, U.S.		Commissioned in October 1996	Bagasse	IGT biomass gasification technology demonstration plant
Sydkraft, Varnamo, Sweden	6 MWe + 9 MWth	In operation since June 1996	Biofuel	Pilot plant; Ahlstrom CFB Gasifier, Sydkraft and Foster Wheeler JV

TABLE 2.12 Summary of Gasification Technologies

Name/Description	Key Features	Experience
	Entrained-Flow Gasifier Technologies	
Babcock Borsig Power (Noell) technology. Also known as the Noell entrained-flow technology. First developed in 1975 in the former East Germany for the gasification of lignite in a 3 MW pilot plant. A full-scale (130 MW) gasifier was built in the 1980s to produce syngas and town gas. The technology was known as the GSP process before being acquired by Noell in 1991.	Process can be a dry-fed or a liquid-fed oxygen-blown slagging gasifier. If solid fuel is to be gasified, it is first pulverized, then pneumatically conveyed to the feeding system and dry-fed together with oxygen and steam through a burner located at the top of the gasifier. Depending on the fuel ash content, the gasification chamber can be covered by either a cooling screen or a cooling wall. Both the refractory and the solid slag provide thermal insulation and maintain the tube surface temperature below 230°C. To allow the solidified slag to regenerate continuously, only fuels with an ash content of more than 1 wt%, such as coals, can be processed in the gasifier lined with a cooling screen. Heat removed by the cooled tube wall represents 2–3% of the total heat produced during gasification and is used to generate low-pressure steam. Syngas saturated with water is further cooled to 150–200°C and recycled to the quench sprays within the gasifier. The	The only Noell gasifiers in commercial operation are at Schwarze Pumpe (Germany) and a newly built one, the BASF Seal Sands located in Middlesbrough in the UK. The BBP Research and Development Centre based at Freiberg, Germany, comprises two facilities with capacities of 5 and 10 MW. The smaller one was originally designed in 1979 for the gasification of both solid and pulverized solid materials. The pilot plant is presently used by Dow Chemical for the development of the gasification of chlorinated wastes. The second one was also designed for the gasification of pulverized materials (coal, waste), liquids, and slurries (waste oil, sludge, paint waste) and built in 1997. A wide range of coals, from anthracites to brown coals, have been gasified in the two pilot plants since the 1980s. BBP claims that it is capable of providing appropriate test conditions to optimize feedstock preparation prior to gasification as well as to determine the optimum gasification

Hitachi EAGLE project. Its objective is the development of an Integrated coal Gasification Fuel Cell (IGFC) combined cycle. The project developed in Japan is sponsored by New Energy Industrial Technology Development Organization (NIDO) and the Agency of Natural Resources and Energy of METI (Japanese Minister of Economy, Trade, and Industry).

bottom part of the gasifier consists of a quench bath that cools and solidifies the slag, which is removed in a granular form.

Coal gasification unit is an oxygen-blown entrained-flow gasifier that can process up to 150 ton coal/day. The gasifier is a water-cooled tube that is lined by a high-temperature-resistant castable. Pulverized coal is pneumatically transported by nitrogen to the gasifier, where it is injected into the gasifier chamber through two types of burners at a pressure of 2.5 MPa. The two sets of burners are installed tangentially to the gasifier sidewall, allowing a spiral flow of coal and oxygen from the upper stage to the lower stage and making particle residence times much longer than those of a gas stream. Enough oxygen is fed to the lower burner to melt the slag. Molten slag solidifies on the gasifier wall as a first layer and subsequent molten slag flows over the layer of the solidified slag to the slag tap hole at the bottom of the gasifier, and it is quenched with water and finally removed via a lock hopper. Coal fed to the upper burners is reacted at lower temperature with a smaller amount of oxygen; it is then gasified and converted to reactive char. The char moves down along the conditions for more than 80 different fuels, including 30 coals.

The raw gas produced together with the fly ash and the remaining char particles go up toward the exit of the gasifier. They enter a syngas cooler, where they are cooled to 450°C prior to going through a cyclone and a filter that retain most of the fly ash and the char particles, which are finally reinjected into the gasifier by pneumatic transport under nitrogen. The syngas goes successively through a water scrubber to remove halides and is desulfurized to be cleaned enough to comply with the strict tolerance limits of fuel cells.

TABLE 2.12 (continued)   Summary of Gasification Technologies

Name/Description	Key Features	Experience
Mitsubishi Heavy Industries (MHI). The air-blown MHI gasifier is divided into two sections: a lower combustion section, which is connected by a diffuser to an upper reducing section.	spiral gas flow and mixes with high-temperature gas in the lower portion of the gasifier, where gasification proceeds further. Dry pulverized coal is fed at two points into the gasifier, with half of the coal being fed into the combustor together with air, where it is burned to produce CO. The temperature inside the combustor is sufficiently high to melt the coal ash without the addition of flux. The slag runs to the bottom of the gasifier, where it is quenched in a water bath and removed using a lock hopper system. The gas produced in the combustor rises to the reducing section, where the remaining coal is added. Coal is then gasified in the reducing section to produce a low CV syngas mainly formed of nitrogen. As the reducer section is at a lower temperature than the combustor section, any molten ash carried upwards is solidified. The syngas produced exits the gasifier through a syngas cooler and then cyclones to collect the chars, as the coal is not completely gasified in the reducing section. Chars collected in the cyclones are then reinjected at the base of the gasifier to ensure	The MHI gasification technology has been tested in Nakoso (Japan) in two pilot-scale gasifiers. A new IGCC project has been started that is a 250 MW air-blown IGCC demonstration plant located in Nakoso, where the former pilot plants were based, and will process up to 1500 ton/day of coal, which is about eight times more than in the former 200 ton/day pilot plant. The system will have a unique feature: the oxidizing gas will be partially extracted from the gas turbine compressor and will be enriched with oxygen coming from an independent air separation unit, making the gasifier operation more stable and giving a certain flexibility to the system that does not exist in the two highly integrated European IGCC plants. An advantage of the MHI two-stage dry-fed entrained-flow gasifier, compared with the one-stage gasifiers, is that the syngas temperature at the outlet of the gasifier is not as high as the one flowing out of a one-stage gasifier and

The only commercial-scale unit is based in Puertollano in Spain (capacity of 338 MWe). It is the largest unit worldwide based on solid fuels. The plant has been operating since 1996 and can process up to 2600 ton/day of coal/petcoke fuel mixed with limestone (2 wt%) and produces 180,000 m³/day of raw gas. The annual production of slag (85% of the ash in weight) and fly ash (15% of the ash in weight) are respectively 120,000 ton and 12,000 ton per year. The demonstration project has now attained commercial development with a gross efficiency of 47.2% (net efficiency of 42%). In April 2002, the company was claiming a variable cost of 12.5 E/MWh compared to 39.5 E/MWh if it had been operating with natural gas.

The first commercial IGCC plant using the Shell Coal Gasification Process is Buggenum in the Netherlands, which was built in 1993. The plant achieves an overall efficiency of 43% that could be increased to over 50% if using

hence does not require a large radiant cooler or a quenching system to mix cold recycled gas with the syngas. The overall cost of the process should then be less than that of existing IGCC plants.

Prenflo (PRessurized ENtrained FLOw) gasification technology.
Initially developed by Krupp Uhde. Krupp and Shell. They split in 1981 and both developed their own coal gasification process.

complete carbon conversion. Because of the very high temperatures reached in the combustion section, this type of gasifier is well suited to gasify the very-high-ash-melting-point Australian coals without any addition of fluxing agent.

Coal is fed together with oxygen and steam through four burners located at the lower part of the gasifier. Syngas is produced at a temperature of 1600°C and is quenched at the gasifier outlet with recycled cleaned gas to reduce its temperature to 800°C. Then the syngas flows up a central distributor pipe and down through evaporator stages before exiting the gasifier at a temperature of 380°C. The raw gas is then dedusted in two ceramic candle filters, and a part of it is recirculated into the syngas cooler. The syngas is finally washed in a Venturi scrubber. Slag formed during the gasification process is quenched in a water bath and is then removed through a lock hopper system.

Shell Coal Gasification Process (SCGP)
Uses single-stage upflow gasifier. Gasifier is a carbon steel vessel that contains a gasification chamber enclosed by a non-refractory membrane wall. Water circulating through the membrane wall is used to control

TABLE 2.12 (continued)   Summary of Gasification Technologies

Name/Description	Key Features	Experience
	the temperature of the gasifier wall and to raise steam. Dried, pulverized coal is stored under nitrogen prior to entering the coal feed system, where it is pressurized and then pneumatically transported into the gasifier. Coal, oxygen, and steam enter the gasifier through horizontally opposed coal burners. The gasifier operates under pressure (2–4 MPa) at 1500°C and more. This ensures ash melting and the formation of a molten slag that runs freely down the membrane wall via the slag tap, to a water-filled compartment at the bottom of the gasifier. The syngas produced is quenched at the gasifier exit with cooled recycled product gas to lower the temperature to around 900°C. The syngas is then further cooled to 300°C in the syngas cooler, raising steam, and is filtered using ceramic filters. About 50% of the cooled syngas is recycled to the top of the gasifier to act as a quenching medium for the gas exiting the gasifier chamber. The rest of the gas is washed to remove halides and NH_3 and then passed to the desulfurization unit. Fly ash leaving the gasifier used to be recycled to the	the most recent gas turbines. The Buggenum design processes coal with natural gas as backup. The plant can process up to 2000 ton/day of fuel. A demonstration plant (220 ton/day) at Oil Deer Park Manufacturing Complex in Houston completed tests proving the ability of the SCGP to gasify more diverse types of coals (220 ton/day of bituminous coals or 365 ton/day of high-moisture, high-ash lignite) before being shut down in 1991. Any coal that can be milled to the right size and pneumatically transported can be gasified in the Shell entrained-flow gasifier. Some adjustments have to be made in order to keep the SCGP performances optimal when changing coal. Bituminous coals require, in most cases, steam injection and oxygen/MAF (moisture- and ash-free) coal ratios from 0.85–1.05 for producing a syngas with a CO/H_2 ratio of 2.2–2.4 and 1–2.5% CO_2. Sub-bituminous coals and lignites normally don't require steam injection and can be operated with oxygen/MAF coal ratio between 0.8 and 0.9, producing syngas with some 3–5% CO_2 and a CO/H_2 ratio of 2.0–2.2. Anthracites

coal-feed system then discharged as slag and sold, but fly ash has proved to have a substantially higher market than the slag and is no longer recirculated into the gasifier.

require a higher oxygen/MAF coal ratio of 1.0–1.1, a higher steam/oxygen ratio of 0.15–0.3, and produce a syngas with similar CO_2 contents as bituminous coal (1–2.5% CO_2, but a higher CO/H_2 ratio of 2.4–2.6). The ash content of a coal has an impact on the performance of the SCGP process in terms of efficiency, as slag forms part of the insulation of the wall of the gasifier and then prevents excessive heat loss during the gasification reaction. Sulcis, a new IGCC project based on the SCGP technology, was being developed in Sardinia, Italy. The project is presently in stand-by due to financial reasons but expected to go ahead. It was planned to have similar characteristics as the Buggenum plant. The Sulcis plant has been designed to gasify 5000 ton/day blends of local coal (high-sulfur, high-ash sub-bituminous coal) and imported LHV coals. A large IGCC demonstration plant is also planned to be built at Yantai Power Plant in Shangong province in China. Technical pre-feasibility studies were carried out in 1994–95. Development prospects were predicted and comparisons were made with CFBC, PFBC-CC, and supercritical units. Two 400 MW IGCC units should be installed. Their net

TABLE 2.12 (continued) Summary of Gasification Technologies

Name/Description	Key Features	Experience
		efficiency is planned to be more than 43%. They are designed to gasify bituminous coals with a high sulfur content (2.5–3%) from Yanzhou in Shangong. Sulphur will be recovered as elemental sulfur, with a predicted removal efficiency of 98%. Three other gasification plants are planned to be developed by Shell in partnership with Sinopec in China, and a fourth one is under feasibility study. The plants will all produce syngas for ammonia/urea production or H_2 for other chemical plants (methanol, oxo), replacing naphta reformers, oil gasifiers, or outdated coal gasifiers.
	Slurry-Fed Gasifier Technologies	
E-Gas (former Destec) Technology	This technology has already been described. Some general characteristics: The gasifier is a two-stage pressure shell lined with refractory material in which coal to be gasified is injected as a preheated slurry. The dry coal concentrations in the slurry range from 50 to 70 wt%, depending on the inherent moisture and quality of the feed. Part of the	Wabash River (U.S.) is the only E-Gas gasifier in operation. Prior to the repowering of the Wabash River plant to an IGCC, some tests of bituminous coals, including high-sulfur coals, were performed in a 2200 ton coal/day plant based in Plaquemine, LA, in the early 1990s. The Wabash River power plant is designed to use a range of local coals with a

maximum sulfur content of up to 5.9% (dry basis) and a higher heating value of 31.4 MJ/kg (moisture- and ash-free). It is presently operating on Illinois No. 6 coal. Alternative fuels (petcoke) have also been successfully tested at Wabash River and future tests may include coal fines. Coal fines are believed to be a promising fuel in the locality of the Wabash River facility, as they are produced by the existing operations of the adjacent mine. They are also available from surface reserves, where the fines have been landfilled in the past and are predicted to be 40–60% cheaper than the present coal delivered to the facility.

coal slurry (80%) is injected with oxygen (95%) through two burners at the lower stage of the gasifier, where it is partially combusted at a temperature of 1350-1400°C and a pressure of 3 MPa. Molten ash formed flows down the gasifier and is removed through a tap hole into a water quench. There is no lock hopper for ash removal in the E-Gas gasifier, and that has the advantage of reducing the overall height of the system. The fuel gas produced in the lower stage flows upwards into the upper stage, where it can react with the remaining 20% coal slurry. This two-stage process presents the advantage of producing a gas with a higher calorific value than the one produced in a one-stage process. The crude gas exiting the gasifier at a temperature of around 1050°C is cooled to 370°C in a firetube syngas cooler. This unit generates saturated high-pressure steam. The firetube syngas cooler is a boiler system with the hot gas circulating on the boiler side, as opposed to a water syngas cooler in which water circulates in tubes in a syngas tank. The firetube is reportedly considerably cheaper than the ones used in the Shell, Texaco, and Prenflo processes. After the cooling step the syngas is cleaned with filters

TABLE 2.12 (continued) Summary of Gasification Technologies

Name/Description	Key Features	Experience
	to remove large ash and char particles that are pneumatically reinjected into the gasifier. The filter elements, made of metal for an acceptable resistance to corrosive syngas, are periodically back-pulsed with high-pressure syngas to remove particulate cake formed on their surface. The particulate cake falls to the bottom of the vessel and is pneumatically transferred together with the high-pressure syngas to the first stage of the gasifier, where it is recycled. Finally, the particulate-free syngas proceeds to the low-temperature heat-recovery system, where it is scrubbed with sour water condensed from the syngas to remove troublesome chlorides and trace elements that could cause corrosion within the piping and vessels as well as form undesirable products in the acid gas-removal system. After scrubbing and reheating, the syngas enters the COS hydrolysis unit, where the COS present in the syngas is converted to H_2S. The syngas is then cooled through a series of shell and tube exchangers to 35°C before entering the acid gas-removal system. This cooling step also condenses water from	

the syngas. Most of the ammonia (NH_3) and some of the carbon dioxide (CO_2) and H_2S present in the syngas are absorbed in the water as dissolved gases.

Texaco Technology

The gasifier is a pressure vessel with a refractory lining, which operates at temperatures in the range 1250–1450°C and pressures of 3 MPa for power generation and up 6–8 MPa for H_2 and chemical synthesis. The feedstocks, oxygen, and steam are introduced through burners at the top of the gasifier. Solid feedstocks such as coal are pre-processed into a slurry by fine grinding and water addition. The slurry is pumped into the burner, and the water that is added with the slurry replaces most of the steam that should normally be injected into the system. Raw gas and molten ash produced during coal gasification flow out towards the bottom of the gasifier. Two alternatives are then available for the recovery of the ash and for cooling the raw gas. The raw gas either can be cooled and cleaned from the slag ash by water quenching or it can be cooled in a radiant syngas cooler from 1400 to 700°C. The heat recovered in the second option is then used to raise steam to be used in the process or for power generation. Molten slag flows

There are several existing projects based on Texaco technology, including the Polk Power Station IGCC project managed by Tampa Electric Company, described earlier. During the first 3 commercial years of operation, ten different coals or coal blends were tested to determine the cheapest feedstock to process while respecting new environmental regulations. The slag removal system of the Polk Power Station was designed for processing coals with a maximum of 12 wt% (dry basis) ash content. The operating temperature of the gasifiers has to be high enough for the coal mineral matter to melt and flow freely down the bottom of the gasifier. Texaco has fixed the minimum heating value of the coals at 30 MJ/kg to produce enough syngas to fully load the combustion turbine. It would be necessary to increase the oxygen supply size as well as the slurry delivery system capacity to be able to run the plant with a lower heating value coal. The plant is designed to accommodate coals with sulfur contents of up to 3.5 wt% (dry basis).

TABLE 2.12 (continued)   Summary of Gasification Technologies

Name/Description	Key Features	Experience
	down the heat-recovery steam generator and is quenched in a batch at the bottom of the cooler and finally removed through a lock hopper system. The quench alternative is the preferred option for coal feedstocks, as they could contain traces of salts (sodium and calcium) that could be corrosive for the syngas coolers at high temperatures. However, this alternative could be a disadvantage for power generation, as thermal efficiency is slightly lowered.	Expensive modifications of the acid gas-removal system were required for higher sulfur content coals than the first base coal (Pittsburgh No. 8-1 with a sulfur content of around 2.5%). Following major problems, the company decided to switch to coal blends with a lower sulfur content. The limit in chlorine concentration in the coals was fixed at 0.15% (dry ash). A higher concentration of chlorine in coals would damage the system. Other coal properties have an influence on the technical and economic aspects of the Texaco-based IGCC operation and necessitate coal testing in the device prior to selecting them for the Polk Power Station. The Texaco technology is also used for chemical plants. Five chemical plants were built after 1993 in the U.S. Eastman Chemicals (Kingsport, U.S.) owns two Texaco quench gasifiers that operate at 70 bar and 1400°C for the production of acetic acid and acetic anhydride. Although the facility is configured for the purpose of making acetyl chemicals, the company claims that gasification and cleanup plants are completely compatible with an electric power

option, and in fact an electric power option of 523 MWe is reported to be under development at Kingsport. This is in line with the new projects of cogeneration of chemicals and electricity sponsored by the U.S. DOE under the Vision 21 program. Another company, Waste Management & Processors, Inc. (U.S.), is presently conducting a techno-economic feasibility study in partnership with Texaco, Sasol, and Nexant for the development of one of the three demonstration Early Entrance Coproduction Plants (EECP) under the Vision 21 program. The objective is the commercialization of a coal gasification/liquefaction technology to produce ultra-clean Fisher-Tropsch transportation fuels with either power, chemicals, or steam as coproducts. The proposed plant location is at the Gilberton Power Plant cogeneration facility in Pennsylvania. It involves the gasification of local waste coals, mainly high-ash-content anthracite wastes derived from an on-site coal cleaning operation that contains coal fines, coal dust, and dirt. Another demonstration EECP project is being developed by Texaco in collaboration with Rentech (Fisher-Tropsch Technology), Brown and Root Services,

TABLE 2.12 (continued)   Summary of Gasification Technologies

Name/Description	Key Features	Experience
		Praxair, and GE Power Systems for the production of electricity and chemicals from coal or petroleum coke. The project involves technical and economic studies of several process options, including syngas composition, Fisher-Tropsch product upgrading, wastewater treatment, catalyst/wax separation, acid gas removal, tail gas utilization, and site selection. There is also a plan for the construction by coal power of a 430 MW IGCC plant based on the Texaco technology near the Hatfield colliery in the North of England. The IGCC project with CO_2 removal and production of H_2 is being studied by Jabobs Consultancy in cooperation with GE. The IGCC power plant is configured to be capable of removing 75% of the feed carbon as CO_2 prior to combustion in the gas turbine. By performing a "sour shift" of the syngas, most of the carbon monoxide should be converted into carbon dioxide and an equal volume of hydrogen. If carbon dioxide removal is performed, then the fuel for the combustion turbine will consist mainly of H_2.

Fluidized-Bed Technologies (Circulating Beds)

BHEL (Indian
Institute of
Technology)

A 168 ton/day coal capacity air-blown pressurized fluidized-bed gasifier IGCC pilot plant (6.2 MWe) was built at Hyderabad, India, following previous gasification tests in an 18 ton/day coal capacity IGCC fluidized-bed gasifier pilot plant and in a 150 ton/day coal moving-bed IGCC pilot plant. The plant consists of a refractory-lined reactor with a 1.4 m inside diameter in the bed, expanding to a 2 m inside diameter at the upper section of the gasifier. Crushed coal (6 mm size or below) is injected into the system via a lock hopper and a rotary coal feeder and then pneumatically transported into the gasifier with a portion of the air used by the plant. The dry granular ash produced during gasification is withdrawn from the bottom of the gasifier through a water-cooled screw extractor and is discharged periodically through an ash lock system. Three refractory cyclones operating in series are used for primary gas cleaning. Fines collected in the first two cyclones can be recycled in the gasifier, but there is also the possibility to collect the cyclone fines, without recycling, through a lock hopper. The gasifier operates at a temperature of 1000°C and pressure of 1.3 MPa to generate a coal gas with a net calorific value of 9.8 MJ/kg.

The 168 ton coal/day demonstration plant was commissioned in 1996 and has since undergone a series of tests in standalone and in IGCC mode, operating for a total of 1200 hours until the year 2000. The plant is designed for the gasification of Indian coals with a high ash content of up to 42%.

TABLE 2.12 (continued) Summary of Gasification Technologies

Name/Description	Key Features	Experience
High Temperature Winkler (HTW)	The HTW process was first developed by Rheinbraun in Germany to gasify lignites for the production of a reducing gas for iron ore. The gasifier consists of a refractory-lined pressure vessel equipped with a water jacket. Feedstocks are pressurized in a lock hopper that is located below the coal storage bin and then pneumatically conveyed to a coal bin. The conveying gas is then filtered and recirculated. Coal in the receiving bin is then dropped via a gravity pipe into the fluidized bed that is formed by particles of ash, semi-coke, and coal. The gasifier is fluidized from the bottom with either air or oxygen/steam, and the temperature of the bed is kept at around 800°C, below the fuel ash fusion temperature. Additional gasification agent is introduced at the freeboard to decompose, at a higher temperature (900–950°C), undesirable by-products formed during gasification. The operating pressure can vary from 1 to 3 MPa, depending on the use of the syngas. The raw syngas produced is passed	A hot gas ceramic candle unit formed of 450 candles was developed and operated for 15,000 hours. The HTW technology manufactured by Rheinbraun was successfully applied for the synthesis of chemicals (methanol) from lignites at Berrenrath, Germany, between 1986 and 1997. The plant was shut down at the end of 1997, as at the time, the process was no longer considered to be economically viable. Another commercial plant has been operating in Finland since 1988, essentially with peat for the production of ammonia. A 140 ton coal/day pressurized HTW gasification plant was also commissioned and built at Wesseling, Germany, in 1989, to supplement research and development of the HTW technology for coal use and particularly to study its future application to an IGCC process for power generation. The plant was designed for a maximum thermal capacity of 36 MW and was operated for 3 years as either an air-blown or an oxygen-blown gasification

plant with pressures up to 2.5 MPa. A wide range of coals was tested in the Wesseling plant, including brown coals and a high-volatile bituminous coal (Pittsburgh No. 8). The Wesseling plant provided the operational data required to design a potential 300 MW commercial IGCC power plant (KoBra), which was finally never built. However, there is presently a project to develop a 400 MW IGCC plant based on the HTW technology (two units) to replace 26 existing Lurgi moving beds at Vresova in the Czech Republic. The new HTW plant (80 ton coal/hour and pressures up to 3 MPa) should operate on Czech lignite and will benefit from years of research and development at the Wesseling and Berrenrath plants. In order to adapt the HTW technology to the Czech lignites and also to the pre-existing Vresova IGCC plant (coal grinding plant, air separation unit, wastewater treatment, and steam turbine), tests were performed by Rheinbraun in an HTW bench-scale gasification unit and compared to results obtained with other coals in the same bench-scale unit and in a demonstration plant.

through a cyclone to remove particulates and then cooled. Solids recovered in the cyclones are reinjected into the gasifier, and dry ash is removed at the bottom via a discharge screw. The syngas cooling system has been the subject of study as to whether to use a water-cooled or a firetube syngas cooler. The main reason was that the existing water-cooled syngas cooler was facing fouling and corrosion problems. A conventional water scrubber system was originally used for gas cleaning but due to blockages, fouling, corrosion, and also the high operating cost of the system, Rheinbraun decided to develop a hot gas filtration system.

TABLE 2.12 (continued)   Summary of Gasification Technologies

Name/Description	Key Features	Experience
Integrated Drying Gasification Combined Cycle (IDGCC)	The IDGCC technology was specifically developed for the gasification of high-moisture low-rank coals by Herman Research Pty Limited in Morwell, Australia. The gasifier is a 5 MW air-blown pressurized fluidized-bed pilot plant that is fed with coal from an integrated drying process. The feed coal is pressurized in a lock hopper system and then fed into the dryer, where it is mixed with the hot gas leaving the gasifier. The heat in the gas is used to dry the coal, while the evaporation of water from the coal cools down the gas without the need of expensive heat exchangers. The gasifier operates at 900°C under 2.5 MPa air pressure. Chars and ash are collected at the bottom of the gasifier and from a ceramic filter and burnt in a separate boiler. The final ash product is similar to that from a conventional low-rank boiler. A wide range of low-rank coals could be processed in the IDGCC, with only small changes in the operating conditions. Coals containing high levels of sulfur can be processed with sorbents, such as limestone or dolomite, directly injected into the bed. This would	As the IDGCC plant is based on a fluidized-bed gasification technology, it is not recommended, as in most of the fluidized-bed technologies, for coals with relatively low reactivities and coals with low ash melting points. As stated by Holt (2001) when looking at environmental considerations and particularly at the concept of CO_2 removal and H_2 production, the IDGCC process, which produces a very moist syngas, can provide the water for the shift reaction without robbing or much reduced robbing of the steam cycle and may have potential for future development. It was reported that the IDGCC process is more efficient and as a consequence more environmentally friendly (lower CO_2 emission) than conventional processes, and would be just slightly less efficient than an Australian black coal IGCC process.

obviate the need for additional cooling of the gas to 40°C for sulfur removal from the very high moisture syngas. The extra cooling would have led to a very large energy loss from water condensation and reduced mass energy for the gas turbine. It is expected that the IDGCC could handle coals with a lower moisture content and a higher ash content.

Kellog Rust Westinghouse (KRW)

The KRW gasifier is an air-blown pressurized system that operates at pressure up to 2 MPa. Coal and limestone are ground and fed together with air and steam at the bottom of the gasifier through lock hoppers. During gasification, sulfur present in coal reacts with crushed limestone to form calcium sulfide, which exits the gasifier along with the residual char and ash and passes to a sulphator. In the sulphator, which is a fluidized-bed combustor, the residual char is burnt and calcium sulfide is oxidized to calcium sulfate, which is considered more environmentally friendly. The heat released by the sulphator is used to raise steam. Raw gas leaves the gasifier at a temperature of around 900°C and passes through a cyclone, in which most of the entrained particulates are retained and recirculated to the gasifier. The gas is cooled to around 600°C and enters

Pinon Pine in Nevada (U.S.) is the only large-scale coal-based IGCC plant (100 MWe) that is using the KRW technology, and it is also the only one that was designed with a 100% hot gas cleanup. The demonstration plant, owned by Sierra Pacific Resources and sponsored by the U.S. DOE, has had numerous problems. The gasifier had 18 start-ups, and all of them failed due to equipment design. Successes in the project included operation of the combined cycle portion of the plant at 98% availability, efficient removal by the hot-gas filter of particulates from the syngas, and production of a good quality syngas for only 30 hours since the first syngas was produced in 1998. Sierra Pacific Resources, which owns the Pinon Pine Power Plant, was going to be sold to WPS Power Development but the sale has been suspended by the state of Nevada, which

TABLE 2.12 (continued) Summary of Gasification Technologies

Name/Description	Key Features	Experience
		placed a moratorium on the sale of power plants in the state.
The Transport Reactor Gasifier	The transport reactor developed by Kellog Brown and Root Inc. at Power System Development Facility (PSDF) in Alabama, U.S., is a demonstration scale circulating fluidized-bed gasifier plant (26.8 to 64 ton coal/day) designed to operate either as a pressurized combustor or as a gasifier. The transport reactor operates at temperatures between 870 and 1000°C and pressure up to 1.5 MPa, with a much higher circulation rate (50 to 100 times the coal feed rate) and higher velocities (from 7 to 17 m/s) than conventional circulating fluidized-bed gasifiers. These differences result in a higher throughput, a better mixing, and higher mass and heat transfer rates in the transport reactor. The transport reactor is midway between a fluidized-bed and an entrained-flow gasifier but is classified as a fluidized-bed gasifier because of the solid recirculation into the	PDSF gasifier is currently under commission and is being tested with several fluxes and coal blends, including three sub-bituminous Powder River Basin coals, three bituminous coals from Alabama, and East Kentucky and Illinois No. 6. The initial tests with bituminous coals were only preliminary and of limited value due to inefficient solid recirculation. The gasifier and filters were reported to be extremely stable, although the coal-feed system had operational problems with fine grind coals, with an overall carbon conversion of more than 90% and synthesis gas heating values over 4 MJ/m^3. After completion of the first tests several modifications were made to the system. A coke breeze-feed system was designed to raise the temperature in the gasifier above 870°C prior to coal feeding in order to prevent tar formation during start-up. The development
	a hot-gas cleaning system, where the remaining particulates are removed. Final traces of sulfur are also removed by reaction with a metal oxide sorbent in a transport reactor.	

gasifier. Coal is ground close to pulverized grind and is pressurized with nitrogen before being injected into the gasifier. The fuel and the sorbent are separately fed through lock hoppers into the mixing zone, where they are mixed with air and steam along with recirculated solids from the standpipe. The gas along with the entrained solids move up from the mixing zone into the riser, which has a slightly smaller diameter and makes two turns prior to entering a disengager. The disengager removes the larger particles by gravity separation, with most of the remaining particles being removed in the following cyclone. Solids collected in the disengager and the cyclone are recycled to the mixing zone through the standpipe and the J-leg. After passing through the cyclone, the syngas flows through the primary gas cooler and a barrier filter (ceramic or sintered metal candles), which removes almost all the dust remaining in the gas stream. The syngas is further cooled in a secondary gas cooler prior to being discharged through a pressure let-down valve to the syngas combustor. The fine char/ash collected in the barrier filter and the char/ash mixture extracted from the reactor are cooled, depressurized in lock hoppers, and

of the PSDF is supported by a pilot-scale transport reactor development unit (TRDU) sponsored by the U.S. DOE under the Vision 21 program, for the coproduction of power and chemicals. The TRDU has been recently modified with the aim of increasing the reactor performance. This includes an increase in the reactor mixing zone length and the replacement of the J-leg loop seal with an L-valve loop seal. These modifications were undertaken with the aim of increasing both solid recirculation rates and solid density in the mixing zone and also to provide adequate carbon content at the bottom of the mixing zone when operating in oxygen mode. It also led to an increase of the solid back mixing in the mixing zone, thereby increasing solid residence time. The pilot-scale gasifier (TRDU) has been operating on nine different coals ranging from bituminous to lignites with a coal-feed rate of around 2 ton/day for over 2000 hours to establish a database of the operation under both air-blown and oxygen-blown modes. Operating temperatures were varied from 815 to 1050°C, depending on the fuel reactivity and the fuel ash propensy to agglomerate. Future tests are planned in the PDSF gasifier with a view to optimizing results.

TABLE 2.12 (continued)   Summary of Gasification Technologies

Name/Description	Key Features	Experience
	then combined prior to their combustion in an atmospheric fluidized bed combustor (AFBC). The AFBC recovers the char/ash carbon heat content as superheated steam. The multiple passes of the coal/char through the gasification zone leads to a high carbon conversion of 95% in the transport reactor.	
	Hybrid Systems	
Air-Blown Gasification Cycle (ABGC)	The ABGC is a hybrid system that was developed at pilot scale (0.5 ton/hour coal capacity) by the former Coal Technology Development Division of British Coal. The gasifier is based on a spouted-bed design and is operated at pressures up to 2.5 MPa and a temperature between 900 and 1000°C. Coal fed in the gasifier produces a gas with a low calorific value of around 3.6 MJ/m^3. Sorbents such as limestone are also injected into the gasifier to retain up to 95% of the sulfur originally present in coal. Syngas is first cleaned in a cyclone, then cooled to around 400°C and cleaned by a ceramic filter to be finally burnt and expanded through a gas	A laboratory at Imperial College of Science Technology and Medicine in London (UK) studied the impact of several coal characteristics on the gasification reactivity of some international traded coals in bench-scale reactors that could mimic the behavior of single coal particles in the ABGC. Coal characteristics studied included coal maceral composition and coal mineral matter composition. CO_2-gasification experiments with coals and their macerals revealed that it was difficult to predict coal gasification reactivity in the ABGC only from coal maceral composition, although prediction of coal pyrolysis reactivity matched quite well

results obtained in the bench-scale reactors. However, it was concluded that the nature and reactivity of the chars depend on a number of factors that include not only the maceral content of the coals but also the conditions of char formation, such as temperature, pressure, residence time, and parent coal. These affect the two main processes that seem to govern the reactivity, which are the deposition of secondary carbon (by the intraparticle decomposition of volatiles) and change in the base char structure (caused by the development of fluidity and escape of volatiles from the melt and its resolidification). Results show that under conditions relevant to the ABGC, vitrinite is a maceral that melts and swells, liptinite also melts but does not swell or agglomerate and loses a large proportion of its mass by pyrolysis, and the third maceral, inertite, does not melt, but only loses a small proportion of its mass under pyrolysis and is unreactive towards the gasification agent, CO_2. The suite of coals tested were rich in vitrinite, and the maceral seemed to dominate the morphological changes that occurred during char formation. The second part was on the influence of mineral matter composition in

turbine. Only 70 to 80% of the fuel is gasified, and partially gasified char and other solid residues (fly ash and sulphided sorbent residues) produced in the gasifier are then transferred to an atmospheric pressure circulating fluidized-bed combustor (CFBC) operating at a temperature of about 1000°C. Heat generated by the combustion of the char supplies a steam cycle used to drive a steam turbine to supplement the electricity generation. The ABGC process is forecast to have an efficiency of about 46–48%. The ABGC technology was later purchased by Mitsui Babcock Energy Limited (MBEL), which produced in collaboration with GEC Alsthom and Scotish Power plc a design of a demonstration plant while being supported by the European Commission under the THERMIE program. A wide range of UK coals and international steam coals were studied for use in the ABGC.

TABLE 2.12 (continued)   Summary of Gasification Technologies

Name/Description	Key Features	Experience
		coals. Experiments consisted of the pyrolysis and CO_2 gasification of two coals, which were first demineralized and then impregnated with different salts in a wire-mesh reactor in which gasification conditions were relevant to the ABGC. Results showed that although mineral matter contents clearly affect coal conversion under pyrolysis and gasification conditions, it was difficult to find systematic patterns for the effect of specific inorganic components on different coals.
	Moving-Bed Gasifiers	
British Gas/Lurgi (BGL)	The BGL gasifier is a double-walled cylindrical reactor surrounded by a steam jacket. It is fed through a lock hopper with lump coal and a flux of limestone, which are discharged into the top of the gasifier as a sequence of batches. A distributor plate slowly rotates to ensure even distribution of the coal over the top of the bed. Coal is dried and devolatilized at the upper zone of the gasifier, the temperature being approximately 450°C at the top of the bed. It is gasified in the middle	This is an oxygen-blown flow slagging gasifier first developed back in the 1970s by British Gas at the Westfield Development Center in Fife, Scotland, to produce SNG from coal.

zone and combusted in the lower zone. The CO_2 produced in the combustion stage flows through the gasification section and acts as a gasifying agent. Oxygen and steam are added towards the bottom of the bed through tuyeres. This results in a high internal temperature within the gasifier (around 2000°C in the combustion zone), which causes the ash to melt. The molten ash is then tapped off and quenched with water to solidify it. Ash melt characteristics influence bed permeability, and fluxes such as limestone need to be added to modify slag flow characteristics. Consequently the counter-current operation of the product gas exits the gasifier at a temperature of 450–500°C. Tars, high boiling point hydrocarbons, and fine particulates released at the top of the gasifier during the devolatilization step are removed in the quench vessel and returned to the gasifier through injection points near the tuyeres, where they are completely gasified. The gas is cooled and cleaned by a water quench and then scrubbed to remove H_2S.

TABLE 2.12 (continued) Summary of Gasification Technologies

Name/Description	Key Features	Experience
Westfield facility (UK)	The Westfield facility in Scotland was first developed by British Gas and was later purchased by Global Energy. This company plans a major refurbishment of the plant to generate 120 MW of power from a mixture of coal, sewage sludge, and MSW. Global Energy has also applied for permission to build a second plant to gasify coal and MSW to produce 400 MW of electricity.	Trial runs of mixed fuels with up to 75% waste have been demonstrated.
Schwarze Pumpe complex (Germany)	A new BGL gasifier was installed at the Schwarze Pumpe complex in Germany to replace two of the Lurgi gasifiers currently in operation. This unit (85 MWe) gasifies MSW and bituminous coal for the production of electricity and methanol and was recently commissioned.	
Kentucky IGCC project (U.S.)	The BGL technology was chosen for a new 540 MWe power generation IGCC project in Kentucky. The project is developed by Kentucky Pioneer Energy and is sponsored by Global Energy and the DOE. The objective of the project is to demonstrate the reliability, availability, and maintainability of a utility-scale IGCC system using a high-sulfur bituminous coal, coal fines, and a pelletized	

refuse-derived fuel (RDF) blend in a BGL gasifier and the operability of a molten carbonate fuel cell (MCFC) fuelled by the desulfurized, heated medium-Btu coal gas produced. Production of Fisher-Tropsch liquids from a portion of the syngas produced is an option that will be also considered. The plant is under construction, and operation is predicted to start in 2004 with first coal as the only feedstock and then coal mixed with an increasing ratio of RDF through the demonstration period. This will allow the development of a database of plant performance at various levels of RDF feed. Global Energy will also finance the construction of a 549 MW gasification plant based on the BGL technology for the cogeneration of H_2 and electricity from coal/MSW mixtures at Lima (U.S.).

Lurgi dry ash gasifier

For many years the process was based on lignite. Further developments were made in the 1950s by Lurgi and Ruhrgas to allow the Lurgi gasifier to process a wider range of feedstock such as bituminous coals. The Lurgi dry ash system was later modified to develop the BGL slagging gasifier. Like the BGL gasifier, it is fed with lump coal and produces tars as a by-product. The major difference

The world's largest concentration of Lurgi gasifiers is in South Africa, where Sasol operates three major complexes. The Sasol plants (Sasol I, II, and III) located in Seconda and Sasolburg gasify approximately 30 million ton/year of bituminous coal to synthesis gas, which is converted to fuels and chemicals via the Fisher-Tropsch process. It was recently announced (International Coal

TABLE 2.12 (continued) Summary of Gasification Technologies

Name/Description	Key Features	Experience
	between the two types of moving beds is that the Lurgi dry ash gasifier uses a much higher ratio of steam to oxygen (4–5:1) than the BGL gasifier (0.5:1), resulting in a much lower bed temperature (1000°C in the combustion zone compared to 2000°C in the BGL) and the production of dry ash. The lower temperature in the dry ash system makes it more suited to reactive coals like lignites rather than bituminous coals. Ash is removed by a revolving grate and is depressurized in a lock hopper. Steam and oxygen are blown up through the grate to provide the oxidant for the gasification process. The gas produced exits the gasifier at a temperature of 300–500°C and is water quenched. The gasifier is surrounded by a water jacket that raises steam, which can be used to provide part of the steam to the gasifier. The Lurgi gasifiers have a maximum diameter of 4 m, which corresponds to an input equivalent to 100 MWe for each gasifier, and as a consequence, these have to be used in series in large power plants.	Report, 2001) that the Sasolburg plant will be converted to natural gas, and natural gas will also be used to supplement coal at the synthetic fuel plant at Seconda from the year 2004. A total of 97 gasifiers (80 at Seconda and 17 at Sasolburg) have a combined production capacity of $3.6 \times 106 \ m^3/h$ pure synthesis gas. Coals used by Sasol can vary substantially in terms of chemical and physical properties, and these variations have a direct impact on the performances of the gasifiers. Blending is used at Sasol in order to keep as close as possible to the characteristics of the ideal coal to process in their gasification plants. The Great Plains synfuel plant owned and operated by Dakota Gasification Co. is the only commercial-scale coal gasification plant in the U.S. that manufactures synthetic natural gas (SNG) (4.7 million m^3/d) and recently anhydrous ammonia (1100 ton/day). The synfuel plant, operating since 1984, is formed of 14 Lurgi dry ash moving-bed gasifiers that process up to 18,000 ton/day of lignite. A flue gas desulfurization unit was added downstream

of the gasifier to allow the plant to process high-sulfur coals. The company has regulated the quality of the lignite used by quantifying the effects of different coal contents on the gasification process, with the aim of improving the plant efficiency as well as to decrease the cost of maintenance of the plant. Coal contents that have been considered as having the largest effect on gasifier operations are sulfur content, sodium content (maintained constant at 6–8%), and fine content (less than 6 mm in size) of not more than 5% in weight. Blending is usually the solution implemented to maintain a constant sodium content but it is secondary for sulfur. A 351 MWe IGCC plant was repowered in 1996 at Vresova in the Czech Republic for the production of steam and electricity. The plant owned by Sokolovska Uhelna, A.S., Vresova, comprises 26 Lurgi dry ash gasifiers and processes lignite with a particle size of 10–30 mm under a pressure of 2.5–2.8 MPa. The plant is producing up to 200,000 m³/h of raw gas, representing only 70% of the total electric plant load. Natural gas is a complementary fuel used predominantly to meet peak load demand. The system is seen as highly flexible. As an example, the coal gas input can be

TABLE 2.12 (continued)   Summary of Gasification Technologies

Name/Description	Key Features	Experience
BHEL's pilot plant	The gasification media, a mixture of air and steam, is fed through a grate that also enables ash removal. A gas cooler is used to recover part of the sensible heat of the gas produced and superheat steam for the gasifier. Further gas cooling as well as tar condensation are done by water quenching. Particulates are removed with a Venturi scrubber. The pilot plant operated for more than 5500 hours (1100 hours as IGCC) with two types of coals having high ash contents, Singareni coal with an ash content of 27–35% and North Karanpura coal with an ash content of 40%. The North Karanpura coal was also tested in the Lurgi pilot-scale plant at the Indian Institute of Chemical Engineering (IICT) under the same gasification conditions. It resulted in a better performance of the BHEL gasifier (calorific	increased from 70% to 100% in 5 minutes. A major disadvantage of the moving-bed technology is the production of large quantities of tars and phenols, a problem that does not exist with fluidized-bed gasifiers. A 6.2 MWe IGCC plant was developed by BHEL at the Trichy unit in India in 1988, as part of a research program for the development of gasification of Indian coals for the production of electricity. The gasification process was based on a moving-bed technology developed in-house after experience on a Lurgi dry ash bed gasifier (pilot-scale 24 ton/day) was gained at the IICT at Hyderabad and at CFRI at Dhanbad. The gasifier is a 2.7 m diameter, 14 m high jacketed moving-bed gasifier with a coal throughput of 150 ton/day. Crushed coal of 5–40 mm size with an ash content of about 35% is the design feedstock for the gasifier, which is operating at 1 MPa pressure.

value and cold gas efficiency) due mainly to the larger scale of the gasifier. However, the availability of the plant was affected by the poor performance of the raw gas cooler due to tar deposition and choking. A direct contact quench was subsequently designed to replace the gas cooler and overcome that problem. The performance of the moving-bed gasifier was also compared to that of a pressurized fluidized-bed gasifier later developed by BHEL at the Trichy unit in Haybedarad in India. Moving-bed gasifiers produce tar-laden gas, which makes the recovery of the sensible heat of the raw gas difficult. They also need coals with a certain particle size (5–30 mm). They produce large effluents containing tars and phenolic acids, requiring elaborate effluent treatment. For these reasons, BHEL decided to develop the fluidized-bed technology for the processing of Indian coals.

3

Biogasification

INTRODUCTION

Biomass can be converted to a synthesis gas by gasification. Biomass syngas consists primarily of carbon monoxide (CO), carbon dioxide (CO_2), and hydrogen (H_2).

The product from gasification has been electricity or heat. Unfortunately, the low value of these products in today's market is insufficient to justify the capital and operating costs for many systems. When gasification is coupled with the production of a higher value liquid fuel, the combination could be a viable alternative energy technology.

OVERVIEW

Biomass is a natural substance, which accumulates solar energy as chemical energy by the process of photosynthesis in the presence of sunlight. Biomass largely contains cellulose, hemi-cellulose, and lignin, having an average composition of $C_6H_{10}O_5$, with slight variations. For the complete combustion of biomass the theoretical amount of air required (defined as the stoichiometric quantity) is 6 to 6.5 kg of air per kg of biomass. The end products are CO_2 and H_2O.

In gasification, biomass is subjected to partial pyrolysis under sub-stoichiometric conditions with the air quantity being limited to 1.5 to 1.8 kg of air per kg of biomass. The resultant mixture of gases produced during gasification process is called biogas, which contains CO and H_2 and is combustible. The raw biogas also contains tar and particulate matter, which have to be removed depending on the application.

The process is best thought of as a thermo-chemical conversion of solid biomass into gaseous fuel. The biogas produced has a low calorific value (1000 to 1200 Kcal/Nm³); however, this product can be combusted at a relatively high efficiency and with good degree of control without emitting smoke. Typical conversion efficiencies of the gasification process are 60 to 70%.

TECHNOLOGY ADVANTAGES

By converting solid biomass into combustible gas the following advantages are captured: The process may be considered a clean technology because of the reduction achieved in CO_2 emissions; the equipment has a relatively small footprint and is compact; there is a high thermal efficiency and a good degree of combustion control; and in areas where biomass sources are readily already available at low prices, gasifier systems offer economic advantages over other energy generating technologies.

Biomass resources are a major component of strategies to mitigate global climate change. Plant growth recycles CO_2 from the atmosphere, and the use of biomass resources for energy and chemicals results in low net emissions of carbon dioxide. Since the emissions of NO_x and SO_x from biomass facilities are also typically low, it is a technology that helps to reduce acid rain.

When applied in regions that favor agricultural and forestry products, the technology helps to develop new markets and serves as a mechanism for rural economic development.

When compared with combustion systems, the fuel gas produced by gasifiers is lower in both volume and temperature than the fully combusted product from a combustor. These characteristics provide an opportunity to clean and condition the fuel gas prior to use. Combustion of the resulting biogas can be more accurately controlled than combustion of the solid biomass. Consequently, the overall emissions from gasification-based power systems (especially NO_x) can be reduced.

GENERAL APPLICATIONS

The biogas can be used for a variety of applications, which include thermal energy for cooking, water boiling, steam generation, and drying; in power applications such as diesel engines (the biogas can be used as a fuel in IC engines for applications such as water pumping); and electricity generation (e.g., biogas can be used in a dual-fuel mode in diesel engines or as the only fuel in spark ignition engines in gas turbines).

For thermal applications, a gasifier can be retrofitted with existing equipment such as ovens, furnaces, and boilers. Thermal energy of the order of 4.5 to 5.0 MJ is released by burning 1 m³ of biogas in the burner. Flame temperatures as high as 1200°C can be obtained by optimal air preheating and pre-mixing of air with gas. This means that the biogas can replace fossil fuels in a wide range of applications. A few of the devices that could be retrofitted with gasifiers include furnaces for melting non-ferrous metals and for heat treatment, tea dryers, ceramic kilns, boilers for process steam, and thermal fluid heaters.

A diesel engine can be operated on dual-fuel mode using biogas. Diesel substitution of over 80% at high loads and 70 to 80% under normal load variations are achievable. The mechanical energy derived can be used either for driving water pumps for irrigation or for coupling with an alternator for electrical power generation. Alternatively, a gas engine can be operated with biogas on 100% gas mode with suitably modified air-to-fuel mixing and an appropriate control system.

COMMERCIAL SYSTEMS

Low- and medium-energy gasifiers have been built and operated using a variety of configurations. Some of the more common configurations include updraft or downdraft fixed beds; moving fluidized beds, where fluidized or entrained solids serve as the bed material; and miscellaneous configurations including moving grate beds and molten salt reactors. Most of these configurations can be designed for operation at either ambient conditions or pressurized conditions up to approximately 20 atm. Gasifier configurations are described by Klass,[1] Quaak et al.,[2] Karlschmitt and Bridgwater,[3] and Stassen.[4]

A gasifier plant consists of a reactor, which receives air and solid fuel and converts them into gas, followed by a cooling and washing train, where impurities are removed. The clean combustible gas at a nearly ambient temperature is available for running diesel-generator sets in dual-fuel mode or gas-engine generator sets suitable for running on biogas alone. In thermal applications, the raw gas cooling and cleaning is limited to the requirements of the thermal process.

In a downdraft reactor configuration the biomass feedstock undergoes drying and devolatilization in the upper zones, which results in the production of char. The volatile matter undergoes oxidation in the combustion zone, with air being partially drawn from the open top and partially supplied by air nozzles located after the devolatilization zone. Special washing and cleaning systems have been developed as an integral part of gasification systems, designed to further reduce the levels of tar and particles in the cold biogas to lower levels.

[1] Klass, D. L., 1998. Biomass for Renewable Energy, Fuels, and Chemicals. Academic Press, San Diego, California. 651 pp.

[2] Quaak, P., H. Knoef, and H. Stassen, 1999. Energy from Biomass. A Review of Combustion and Gasification Technologies. World Bank Technical Paper No. 422, Energy Series, The World Bank, Washington D.C. 78 pp.

[3] Karlschmitt, M., and A.V. Bridgwater, eds., 1997. Biomass Gasification and Pyrolysis. ISBM 1872691714. CPL Press, Newbury UK. 550 pp.

[4] Stassen, H.E.M., 1995. Small-Scale Biomass Gasifiers for Heat and Power: A Global Review. World Bank Technical Paper No. 296, World Bank, Washington D.C.

For either low- or medium-energy gasifiers, the selection of a particular reactor design directly influences the characteristics of the raw product gas, including its temperature and the amounts of tars and particulates present. It is important to carefully select gasifier designs that will match product characteristics with end uses. To date, most gasifiers have been designed with either fixed-bed or fluidized-bed technologies.

The influences of various gasifier types on the product composition are summarized in Figure 3.1. A brief description of each type of gasifier is provided in Table 3.1.

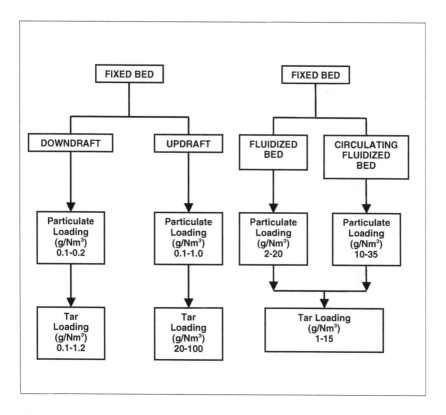

Figure 3.1 Ranges for particulate and tar levels for different biomass gasifiers. (After Neeft, J.P.A., Knoef, H.A.M., and Onaji, P., 1999. Behavior of Tars in Biomass Gasification Systems. NOVEM EWAB Program Report 9919. Available from MHP Management Services, P.O. box 127, 3950 AC Maarn, Netherlands. 75 pp.)

TABLE 3.1 Summary of characteristics of biomass gasifiers

Configuration	Description	Operational Characteristics
Fixed Bed	These have a stationary reaction zone typically supported by a grate and are usually fed from the top of the reactor. They can be designed in either downdraft or updraft configurations.	Fixed-bed gasifiers are relatively easy to design and operate and they are best suited to small- to medium-scale applications with thermal requirements up to a few MW. At larger scale, fixed-bed gasifiers can encounter problems with bridging of the biomass feedstock. This leads to uneven gas flow. Achieving uniform temperatures throughout the gasifier at large scale can also be difficult due to the absence of mixing in the reaction zone. Most fixed-bed gasifiers are air-blown and produce low-energy gases, although oxygen-blown designs have been tested.
Fixed-Bed Downdraft	With fixed-bed downdraft gasifiers, air or oxygen passes downward through the gasifier, co-current to solids flow. Gasification occurs near the bottom of the gasifier in a shallow hot reaction zone composed of a layer of charcoal. The oxidation of the biomass generates heat to sustain the gasification process.	Most tars in the product gas are destroyed by thermal cracking as they pass through the hot reaction zone, and the raw product then exits the gasifier at high temperature. The particulate levels in the product gas are typically low due to the absence of turbulence in the gasifier, but the gas may contain alkali vapors as it exits the hot reaction zone.
Fixed-Bed Updraft	With fixed-bed updraft gasifiers, the air or oxygen passes upward through a hot reactive zone near the bottom of the gasifier in a direction countercurrent to the flow of solid material. Exothermic reactions between air/oxygen and the	Heat in the raw gas is transferred to the biomass feedstock as the hot gases pass upward, and biomass descending through the gasifier undergoes drying, pyrolysis, and then gasification. The product gas exits at a low temperature. Tar concentrations in the product gas are high since the tar vapors formed by

pyrolysis reactions are swept upward through the reactor with the product gas. Particulate levels in the raw product gas are low because of the non-turbulent conditions. Ash is swept along with the solids in the opposite direction of the gas flow and is withdrawn from the bottom of the gasifier.

charcoal in the bed drive the gasification process. Most smaller-scale gasifiers are based on fixed-bed designs.

Fluidized Bed

Fluidized-bed gasifiers are considered to be those systems with a moving bed of inert material such as sand. Examples include bubbling fluidized-bed gasifiers, where the bed material is agitated by gases flowing through it; circulating fluidized bed gasifiers, where the bed material circulates between the gasifier and a secondary vessel; and entrained-bed gasifiers, where the solids are entrained in the gas flow at higher velocities. The gas used to fluidize or entrain the bed material can be air, oxygen, steam, recycled product gas, or a combination. Fluidized-bed gasifiers can be sized for medium- to large-scale facilities and are best suited to situations where there is a relatively constant demand for the gas product. Essentially, all of the larger-scale gasifier facilities built and tested in the last decade use fluidized-bed designs. The product gas exiting from the

Depending on the specific design, biomass can be fed into the top, bottom, or middle of the moving bed. Heat to drive the gasification reaction can be provided in a variety of ways in fluidized-bed gasifiers. Direct heating occurs when air or oxygen in fluidizing gas partially oxidizes the biomass and heat is released by the exothermic reactions that occur. Indirect heating methods such as internal heat exchangers, using preheated bed material, or other means can also be used to drive the gasification reactions. The turbulence in the bed promotes effective mixing and efficient heat transfer. The raw product gas exits the reactor at high temperatures. Since the gas is typically hot, it may also contain vaporized alkali salts. The amount of tars in the raw gas can vary significantly depending on the specific reactor design. Typical moving-bed gasifiers produce lower concentrations of tar than fixed-bed updraft types, but more than fixed-bed downdraft reactors. Depending on the reactor design, the composition of the tars can range from partially cracked pyrolysis products to highly refractory polyaromatic hydrocarbons.

TABLE 3.1 (continued) Summary of characteristics of biomass gasifiers

Configuration	Description	Operational Characteristics
	gasifier typically has high levels of particulates as a result of the turbulence in the reactor. The particulate matter consists both of ash originating from the biomass and fine particles of attrited bed material.	
Miscellaneous	Moving grate gasifiers similar to some combustors and incinerators have stemmed from fixed-bed downdraft designs. The lower fixed grate is replaced by a moving or rotating grate or platform. Other designs such as molten salt gasifiers and supercritical fluid gasifiers have been proposed.	The temperature of the product gas and its composition will vary depending on the type of design. Characteristics generally fall within the limits of the fixed- and moving-bed designs described above.

CONTAMINANTS

The biogas generated from gasification will generally contain contaminants that require removal. The principle contaminant classes encountered are particulates, alkali compounds, tars, nitrogen-containing components, sulfur, and low-molecular-weight hydrocarbons (e.g., methane and ethane). The hydrocarbons are beneficial to fuel gases since they tend to increase the heating value of the product. However, they are undesirable in synthesis gases, where they are unreactive and reduce overall conversion efficiencies to the desired product. Excess hydrocarbons may also deactivate catalysts in the system.

Particulates include solid-phase materials entrained in the raw product gas as it exits the gasifier. They include the inorganic ash that is associated with mineral matter in the biomass feedstock, unconverted biomass in the form of char, or material from the gasifier bed.

Particulate matter is generated in large-scale gasifiers, which rely on bubbling or circulating fluidized configurations to ensure uniform bed conditions during gasification. Such configurations support turbulent conditions that result in high particulate loadings in the product gases. Turbulent-flow fluidized-bed systems therefore include cyclones to separate the bed material from the product gases. The cyclone serves as the initial particulate removal technology and captures the bulk of coarse particulates. However, finer fly ash will remain in the gas stream. The resulting fly ash consists of smaller-diameter material that can create operational and visual emissions problems if not removed.

A source of fly ash particulates is the mineral matter in the biomass feedstock. As material is gasified, the inorganic matter from the feedstock may be either retained in the gasifier bed or entrained in the product gas and swept out from the reactor. The mineral concentrations in clean wood are typically 1 to 2%, and herbaceous crops may contain up to 10% or more. Crop residues such as straw or rice hulls typically contain 15 to 20% inorganic material. Mineral matter from other sources may also contribute to ash formation that is entrained in the biogas.

A potential source of particulates is char formed when the biomass feedstock is incompletely gasified. These particles undergo devolatilization and subsequent reactions at gasification temperatures that leave them less reactive than fresh biomass feedstock. They can pass through the gasifier before they are completely gasified, particularly in those configurations with turbulent beds. Char entrained in the product gas also represents unconverted biomass that contributes to lower conversion efficiencies.

Generally, large-scale gasifiers are capable of achieving 98 to 99% carbon conversion efficiencies. Hence, 1 to 2% of the carbon in the feedstock remains as solids. Collection of this material and subsequent re-injection of the char into the gasifier can increase overall gasification efficiencies. It must be recognized that particulates are undesirable, especially those that are mineral in nature. These can abrade and damage downstream equipment.

Biomass feedstocks may contain significant amounts of alkali salts, particularly potassium. Potassium is an element required for plant growth, and concentrations are particularly high in rapid-growth biomass such as grasses or related energy crops. The high alkali content of some biomass feedstocks can create gas cleanup problems. Eutectic sodium and potassium salts in the ash material can vaporize at moderate temperatures of about 700°C. Unlike the solid particulates that can be separated by physical means such as fabric filters, the vaporized alkali compounds tend to remain in the product gas at high temperature. As such, the alkali vapors cannot be easily removed from the hot gas stream by simple mechanical cleanup technologies. Condensation of the vaporized alkali typically begins at about 650°C on particles in the gas stream, with subsequent deposition on cooler surfaces in the system such as heat exchangers, turbine expansion blades, or similar areas.

The problem of alkali deposition in biomass-fueled turbine systems can be a major problem with direct combustion gas turbine systems. A glassy ash material can deposit on the blades of the expansion turbine at inlet temperatures above about 785°C. This problem arises from alkali vapors that pass through gas cleanup systems and deposit as the gases cool during expansion. The subject of alkali vaporization and subsequent ash deposition phenomena are discussed by Miles et al.,[5] Jenkins et al.,[6] Baxter et al.,[7,8,9] and Dayton.[10]

The need to remove alkali material from the biogas stream depends on the end use for the gas. Problems associated with alkali vapor formation and deposition are critical in systems where the hot biogas is to be used without significant cool-down. Moderate gas cooling followed by removal of bulk particulates provides adequate cleaning for simple boiler systems that can tolerate some ash deposition. Other applications such as gas turbines require cleaner fuel gases. Since turbines operate at high rotational speeds, deposition can

---

[5] Miles, T. R., T.R. Miles Jr., L.L. Baxter, R.W. Bryers, B.M. Jenkens, and L.L. Oden, 1996b. Alkali Deposits found in Biomass Power Plants. Vol. II. NREL/TP-433-8142. National Renewable Energy Laboratory, Golden, Colorado.

[6] Jenkins, B.M., L.L. Baxter, T.R. Miles, Jr., and T.R. Miles, 1998. Combustion Properties of Biomass. Fuel Processing Technology, Vol. 54, Elsevier Science, Amsterdam. pp 17-46.

[7] Baxter, L.L., T Gale, S. Sinquefield, and G. Sclippa, 1997a. Influence of Ash Deposit Chemistry and Structure on Deposit Physical and Transport Properties. Developments in Thermochemical Biomass Conversion, A.V. Bridgwater and D.G.B. Boocock, eds., Blackie. Academic and Professional Press, London. pp. 1247-1262.

[8] Baxter, L.L., T.R. Miles, T.R. Miles, Jr., B.M. Jenkins, T.A. Milne, D. Dayton, R.W. Bryers, and L.L. Oden, 1997b. The Behavior of Inorganic Material in Biomass-Fired Boilers: An Overview of the Alkali Deposits Project. Developments in Thermochemical Biomass Conversion, A.V. Bridgwater and D.G.B. Boocock, eds. Blackie Academic and Professional Press, London. pp. 1424-1446.

[9] Baxter, L.L., 1992. Ash Deposition During Biomass and Coal Combustion: A Mechanistic Approach. SAND92-8216. Sandia National Laboratory, Albuquerque, New Mexico.

[10] Dayton, D., R. French, and T. Milne, 1994. Identification of Gas Phase Alkali Species Released During Biomass Combustion. Proceedings of Bioenergy .94, U. S. Department of Energy Western Regional Biomass Energy Program, Golden, Colorado. pp. 607-614.

cause rapid imbalance and catastrophic failure. For these systems, thorough alkali removal of fine particles or aerosols of alkali salts is necessary.

The presence of alkali salts may create other problems even in systems where deposition of hot vapors is not an issue. Alkali salts can be corrosive to metal surfaces and can poison catalysts such as those in tar cracking and synthesis gas applications.

FORMATION OF TARS

Tar refers to a range of oxygenated organic constituents that are produced by the partial reaction of the biomass feedstock. These materials can be found in the hot gas stream as vaporized material or as persistent aerosols. They tend to condense at cooler temperatures.

Tars include a variety of oxygenated aromatics formed in the pyrolysis step of the gasification process. Generally, tar is considered to be organic molecules with molecular weights greater than that of benzene (benzene's molecular weight is 78). The actual composition of the tar is complex and dependent on the severity of reaction conditions, including gasification temperature and residence time in the reactor.

Tar formation is thought to occur under the following conditions: As the biomass feedstock is heated, it dehydrates and then volatilizes as it thermally decomposes. The volatilized material either can undergo further decomposition to form permanent gases, or it can undergo dehydration, condensation, and polymerization reactions that result in tar formation.

The amount of tar will also vary significantly depending on the gasifier design. Large-scale, turbulent-bed gasifiers typically have tar concentrations in the range of 1 to 15 g/Nm^3, whereas fixed-bed downdraft reactor designs typically produce tar concentrations of 20 to 100 g/Nm^3. Fixed-bed downdraft gasifiers have the potential to produce very low levels of tars, less than 0.1 g/Nm^3, but typically produce concentrations of 0.5 to 1.5 g/Nm^3.

Tars in the biogas can be tolerated in some systems where the gas is used as a fuel in closely coupled applications such as burners. In these situations, cooling and condensation can be avoided, and the energy content of the tars adds to the calorific value of the fuel.

In more demanding applications, tars in the raw product gases can create handling and disposal problems. Tars condense on cool components downstream from the gasifier, resulting in plugging and fouling of pipes, tubes, and other equipment. In temperature regions above about 400°C, tars can undergo subsequent dehydration reactions to form solid char and coke that further fouls and plugs systems. Also, tar formation presents a cleanup problem and costly disposal of a hazardous waste.

The presence of tars in the product gas is highly undesirable in synthesis gas for hydrogen applications. Tar formation represents a reduction in gasification efficiency since less of the biomass is converted to a fuel or synthesis gas. More importantly, tars would degrade the performance of those systems. Tars can deactivate reforming catalysts, and fuel cell toleration of tars is low.

AMMONIA FORMATION

Ammonia is formed from the protein and other nitrogen-containing components in the biomass feedstock. High-protein feedstocks such as animal wastes or alfalfa result in greater ammonia production. Ammonia production is also higher in pressurized gasifiers due to equilibrium considerations and in pyrolytic (rather than air- or oxygen-blown units) due to the reducing environment in those gasifiers.

Acceptable levels of ammonia in the gas stream are dictated by local regulations. Gasification systems must meet emissions regulations; however, the concentrations of ammonia are relatively low with common feedstocks in most gasifiers.

Ammonia in the product stream is undesirable because it can result in the formation of NOx emissions when the product gas is burned. Cleanup of the ammonia is therefore required for systems in locations with strict NOx regulations.

NO$_x$ FORMATION

NO$_x$ is also produced in some gasifiers. It is generally not present in high enough concentrations to create problems. NO$_x$ is produced by the reaction of nitrogen or nitrogen-containing molecules with oxygen at elevated temperatures in the league of those in combustion systems. The lower temperatures in gasification and the nature of the reactive environment limit NO$_x$ production.

Although the gasifier product itself has low levels of NO$_x$, the total systems emissions of this product must be carefully scrutinized. When clean biogas is eventually burned, NO$_x$ will be produced, as it is in most combustion systems with all fuels. The use of biogas rather than solid biomass fuels provides the opportunity to better control the combustion process, which can potentially result in lower NO$_x$ emissions. As such, gasification offers potential environmental emissions advantages over combustion alternatives. However, NO$_x$ may still occur as the gas is burned, and appropriate NO$_x$ control technologies may be needed.

SULFUR

Most biomass feedstocks contain low percentages of sulfur. In contrast, this is a major concern for coal gasification.

Sulfur in the biomass feedstock can be converted to hydrogen sulfide or sulfur oxides during gasification. Wood typically contains less than 0.1% sulfur by weight, and herbaceous crops may contain 0.3 to 0.4%. Some feedstocks such as refuse-derived fuel (RDF) may contain 1% or more, approximately the same as bituminous coal. As a result of the low levels of sulfur in the biomass, the concentrations of H$_2$S and SOx levels in the product gases are below those requiring cleanup in most applications.

The low concentration of sulfur in biomass offers potential advantages for some applications. In cofiring applications, for example, the cleaner combustion gases from biomass dilute those from coal, and the overall concentrations of sulfur per unit of combustion gas are reduced. In most applications where the biomass product is used as a fuel gas, technologies to remove sulfur will not be needed.

Sulfur is a potential problem even at low levels for synthesis gas systems using certain types of catalysts. The production of methanol from synthesis gas, for example, uses catalysts that are poisoned by sulfur. Some tar cracking catalysts are also sulfur sensitive. In those systems, thorough removal of sulfur will be required. Fuel cell systems are also sulfur sensitive.

HYDROGEN PRODUCTION FROM BIOMASS

Approximately 95% of the hydrogen produced today comes from carbonaceous raw material, primarily fossil in origin. Only a fraction of this hydrogen is currently used for energy purposes; the bulk serves as a chemical feedstock for petrochemical, food, electronics, and metallurgical processing industries.

Hydrogen's contribution to the energy market is increasing with the implementation of fuel cell systems and the growing demand for zero-emission fuels. As such, hydrogen production will need to keep pace with this growing market. In the near term, increased production will likely be met by conventional technologies, such as natural gas reforming. In these processes, the carbon is converted to CO_2 and released to the atmosphere.

With the growing concern about global climate change, alternatives to the atmospheric release of CO_2 are being studied. Sequestration of the CO_2 is an option that could provide a viable near-term solution.

Reducing the demand on fossil resources remains a concern throughout the world. Renewable-based processes like solar- or wind-driven electrolysis and photobiological water splitting hold promise for clean hydrogen production. These technologies, however, face significant technical challenges before they become economically viable. Hence, for the near- and mid-term, generating hydrogen from biomass may be the more practical and viable, renewable, and potentially carbon-neutral (or even carbon-negative in conjunction with sequestration) option.

Biomass feedstocks vary greatly in both composition and form. Both moisture and energy content are key parameters. Since biomass is low in density, the transportation costs for both the feedstock and the hydrogen must be balanced with the savings from employing economy of scale. The distribution of hydrogen production sites requires a decision on the transport of both the biomass and the hydrogen. These characteristics make it difficult to compete with natural gas steam reforming without credits.

Since biomass is renewable and consumes atmospheric CO_2 during growth, it can have a small net CO_2 impact compared to fossil fuels. The yield of hydrogen is low from biomass, since the hydrogen content in biomass is low to begin with (approximately 6% versus 25% for methane), and the energy content is low due to the 40% oxygen content of biomass. Since over half of the hydrogen from biomass comes from splitting water in the steam reforming reaction, the energy content of the feedstock is an inherent limitation of the process.

The low yield of hydrogen on a weight basis is, however, misleading since energy conversion efficiency is high. As an example, the steam reforming of bio-oil at 825°C with a five-fold excess of steam has an energy efficiency of 56%. On the other hand, the cost for growing, harvesting, and transporting biomass is high. As such, even with reasonable energy efficiencies, it is not presently economically competitive with natural gas steam reforming for standalone hydrogen production without the advantage of high-value co-products.

It must also be recognized that as with all sources of hydrogen, production from biomass will require appropriate hydrogen storage and utilization systems to be developed and deployed.

Biomass conversion technologies can be divided into direct production technology routes and technologies aimed at the conversion of storable intermediates. Direct routes have the advantage of simplicity. Indirect routes have additional production steps, but have an advantage in that there can be distributed production of the intermediates, minimizing the transportation costs of the biomass. Intermediates can be shipped to a central, larger-scale hydrogen production facility. Both classes have thermochemical and biological routes.

A third area of hydrogen from biomass is metabolic processing to split water via photosynthesis or to perform the shift reaction by photo biological organisms. This is an area of long-term research.

Direct production of hydrogen from gasification is the simplest route. Gasification is a two-step process in which the solid feedstock is thermochemically converted to a low- or medium-energy-content gas. Natural gas contains 35 MJ/Nm^3. Air-blown biomass gasification results in approximately 5 MJ/m^3; oxygen-blown in 15 MJ/m^3.

In the first reaction, pyrolysis, the dissociated and volatile components of the fuel are vaporized at temperatures as low as 600°C (1100°F). Included in the volatile vapors are hydrocarbon gases, hydrogen, carbon monoxide, carbon dioxide, tar, and water vapor. Because biomass fuels tend to have more volatile components (70 to 86% on a dry basis) than coal (30%), pyrolysis plays a larger role in biomass gasification than in coal gasification.

Gas phase thermal cracking of the volatiles occurs, reducing the levels of tar. Char (fixed carbon) and ash are the pyrolysis byproducts that are not vaporized. In the second step, the char is gasified through reactions with oxygen, steam, and hydrogen. Some of the unburned char may be combusted to release the heat needed for the endothermic pyrolysis reactions.

Gasification coupled with water-gas shift is the most widely practiced process route for biomass to hydrogen. Thermal, steam, and partial oxidation gasification technologies are under development. Feedstocks include both dedicated crops and agricultural and forest product residues of hardwood, softwood, and herbaceous species.

Thermal gasification is essentially high-severity pyrolysis, although steam is generally present. An example of this is the Sylvagas (BCL/FERCO) low-pressure, indirectly heated circulating fluid bed. By including oxygen in the reaction gas the separate supply of energy is not required, but the product gas is diluted with carbon dioxide and, if air is used to provide the oxygen, then nitrogen is also present. Examples of this are the GTI (formerly IGT) High-Pressure Oxygen-Blown Gasifier, as well as the circulating fluid bed by TPS Termiska.

Other relevant gasifier types are bubbling fluid beds being tested by Enerkem and the high-pressure high-temperature slurry-fed entrained flow Texaco gasifier.

All of these gasifier examples will need to include significant gas conditioning, including the removal of tars and inorganic impurities and the conversion of CO to H_2 by the water-gas shift reaction:

$$CO + H_2O \rightarrow CO_2 + H_2$$

There is great interest in the conversion of wet feedstocks by high-pressure aqueous systems. This includes the supercritical-gasification-in-water approach as well as the supercritical partial oxidation approach by General Atomics. Pyrolysis to hydrogen and carbon is being explored as a viable technology for carbon sequestration, although most work is applied to natural gas pyrolysis. Biomass or biomass-derived intermediates could be processed in this way.

Biological conversion via anaerobic digestion is currently being practiced, resulting in methane that can be processed to hydrogen by conventional steam reforming processes.

Pyrolysis produces a liquid product called bio-oil, which is the basis of several processes for the development of fuels, chemicals, and materials. The reaction is endothermic:

$$\text{Biomass} + \text{Energy} \rightarrow \text{Bio-oil} + \text{Char} + \text{Gas}$$

The oil can be formed in 66 wt% yields. Catalytic steam reforming of bio-oil at 750 to 850°C over a nickel-based catalyst is a two-step process that includes the shift reaction:

$$\text{Bio-oil} + H_2O \rightarrow CO + H_2$$

$$CO + H_2O \rightarrow CO_2 + H_2$$

The overall stoichiometry gives a maximum yield of 17.2 g H/100 g bio-oil (11.2% based on wood).

Regional networks of pyrolysis plants could be established to provide oil to a central steam reforming facility. The process is compatible with other organic waste streams, such as aqueous-steam fractionation processes used for ethanol production and trap grease. Methanol and ethanol can also be produced from biomass by a variety of technologies and used for on-board reforming for transportation. Methane from anaerobic digestion could be reformed along with natural gas. Methane could be pyrolyzed to hydrogen and carbon if markets for carbon black were available.

The U-Gas® process for producing hydrogen from coal has been developed by IGT from over 50 years of coal-conversion research. It comprises a single-stage, non-slagging, fluidized-bed gasifier using oxygen or air. In general, conventional gasification of biomass and wastes has been employed with the goal of maximizing hydrogen production. Various investigators have studied biomass and coal catalytic gasification for hydrogen and methane. See for example Hauserman and Timpe[11] and the Recommended Resources section at the end of this chapter.

Pacific Northwest Laboratories studied the gasification of biomass to produce a variety of gaseous fuels by use of appropriate catalysts. Cox et al.[12] describes an approach to thermochemical gasification of biomass to hydrogen. The process is based on catalytic steam gasification of biomass with concurrent separation of hydrogen in a membrane reactor that employs a permselective membrane to separate the hydrogen as it is produced. The process is particularly well suited for wet biomass and may be conducted at temperatures as low as 300°C. One study was conducted at 4000 psi and 450°C, though most were at 15 to 30 psi. The process was named SepRx. Optimal gasification conditions were found to be about 500°C, atmospheric pressure, and a steam/biomass

---

[11] Hauserman, W.B. and Timpe, R.C. (1992). Catalytic Gasification of Wood for Hydrogen and Methane Production. *Energy & Environmental Research Center,* University of North Dakota:pp. 1-38.

[12] Cox, J. L.; Tonkovich, A. Y.; Elliott, D. C.; Baker, E. G., and Hoffman, E. J. (1995). Hydrogen from Biomass: A Fresh Approach. *Proceedings of the Second Biomass Conference of the Americas,* (NREL/CP-200-8098; CONF-9508104) August, Portland, Oregon:pp. 657-675.

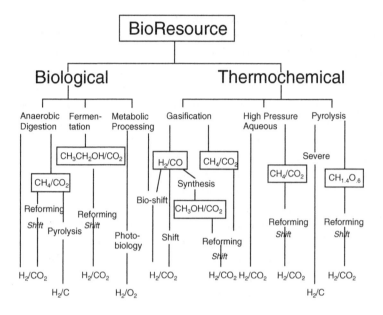

Figure 3.2 Summary of hydrogen production routes from biomass. (Source: Milne, T.A. et al. *Hydrogen from Biomass: State of the Art and Research Challenges*, NREL, IEA/H2/TR-02/001).

ratio equal to 10/1. In the presence of a nickel catalyst, hydrogen at 65% (volume) was produced under these conditions.

A study of almond shell steam gasification in a fluidized bed revealed that, over the range 500 to 800°C, smaller particle size yielded more hydrogen than did higher temperatures (see Rapagna and Foscolo[13]). In this work catalytic steam gasification of biomass was studied in a bench-scale plant containing a fluidized-bed gasifier and a secondary fixed-bed catalytic reactor. The catalytic converter, using different steam reforming nickel catalysts and dolomite, was tested over a range of 660 to 830°C. Fresh catalyst at the highest temperature yielded 60% by volume of hydrogen.

Figure 3.2 provides a summary of the technology routes under development.

---

[13] Rapagna, S., and Foscolo, P. U. (1998). Catalytic Gasification of Biomass to Produce Hydrogen Rich Gas. *Int. J. Hydrogen Energy*; 23, (7):pp. 551-557.

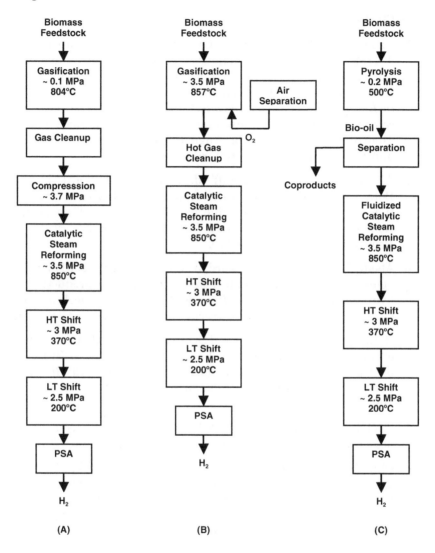

Figure 3.3 Direct gasification routes to hydrogen conversion from biomass.

Figure 3.3 provides the reader with a summary of the leading direct gasification routes of biomass to hydrogen conversion.

RECOMMENDED RESOURCES

1. Aznar, M.P., Caballero, M.A., Gil, J., Olivares, A., and Corella, J. (1997). Hydrogen by Biomass Gasification with Steam-O_2 Mixtures and Steam Reforming and Co-Shift Catalytic Beds Downstream of the Gasifier, *Making a Business from Biomass in Energy, Environment Chemicals, Fibers and Materials, Proceedings of the 3rd Biomass Conference of the Americas, Vol. I,* Montreal, Quebec, Canada, August 24-29, 1997, R.P. Overend and E. Chornet (eds). pp. 859–860.

2. Abedi, J., Yeboah, Y.D., Realff, M., McGee, D., Howard, J., and Bota, K.B. (2001) An Integrated Approach to Hydrogen Production from Agricultural Residue for Use in Urban Transportation, *Proceedings of the 2001 U.S. DOE Hydrogen Program Review* (NREL/CP 570-30535).

3. Abedi, J., Yeboah, Y.D., Howard, J., and Bota, K.B. (2001) Development of a catalytic fluid bed steam reformer for production of hydrogen from biomass, *5th Biomass Conference of the Americas, Orlando, FL (Cancelled).* Abstracts to be published on CD/ROM.

4. Bakhshi, N.N., Dalai, A.K., and Srinivas, S.T. (1999). Conversion of Various Biomass-derived Chars to Hydrogen/High Btu Gas by Gasification with Steam, *Proceedings of the 4th Biomass Conference of the Americas, Oakland, CA, August 29–September 2, 1999.* pp. 985–990.

5. Caglar, A. and Demirbas, A. (2001) Hydrogen Rich Gaseous Products from Tea Waste by Pyrolysis, *Energy Sources 23.* pp. 739–746.

6. Chaudhari, S.T., Ferdous, D., Dalai, A.K., Bej, S.K., Thring, R.W., and Bakhshi, N.N. (2000). Pyrolysis and Steam Gasification of Westvaco Kraft Lignin for the Production of Hydrogen and Medium Btu Gas, *Abstracts: Progress in Thermochemical Biomass Conversion, Tyrol, Austria, 17–22 September.*

7. Cox, J.L., Tonkovich, A.Y., Elliott, D.C., Baker, E.G., and Hoffman, E.J. (1995). Hydrogen from Biomass: A Fresh Approach, *Proceedings of the Second Biomass Conference of the Americas* (NREL/CP-200-8098; CONF-9508104) August, Portland, OR. pp. 657–675.

8. Demirbas, A., Karshoglu, S., and Ayas, A. (1996). Hydrogen Resources Conversion of Black Liquor to Hydrogen Rich Gaseous Products, *Fuel Science and Technology International* 14(3). pp. 451–463.

9. Demirbas, A. and Caglar, A. (1998). Catalytic Steam Reforming of Biomass and Heavy Oil Residues to Hydrogen, *Energy Edu. Sci. Technol* 1. pp. 45–52.

10. Ferdous, D., Dalai, A.K., Bej, S.K., Thring, R.W., and Bakhshi, N.N. (2001). Production of H_2 and Medium Btu Gas via Pyrolysis of Lignins in a Fixed-Bed Reactor, *Fuel Processing Technology* 70(1). pp. 9–26.

11. García, L., Sánchez, J.L., Salvador, M.L., Bilbao, R., and Arauzo J. (1996). Hydrogen-rich Gas from Steam Gasification of Biomass Using Coprecipitated Nickel-alumina Catalysts, *Bioenergy 96. The Seventh National Bioenergy Conference*. pp. 859–865.

12. García, L., Salvador, M.L., Arauzo J., and Bilbao, R. (1997). Steam Gasification of Biomass in Fluidized Bed Using a Ni-Al Coprecipitated Catalyst, *Making a Business from Biomass in Energy, Environment Chemicals, Fibers and Materials, Proceedings of the 3rd Biomass Conference of the Americas, Vol. I*, Montreal, Quebec, Canada, August 24–29, R.P. Overend and E. Chornet (eds). pp. 373–382.

13. Hauserman, W.B. (1997). Relating Catalytic Coal or Biomass Gasification Mechanisms to Plant Capital Cost Components, *Int. J. Hydrogen Energy*; 22, (4). pp. 409–414.

14. Hayashi, J.I., Tomioka, Y., Shimada, T., Takahashi, H., Kumagai, H., Chiba, T. (1998). Rapid Steam Reforming of Volatiles from Flash Pyrolysis of Coal. *Hydrogen Energy Progress XII, Proceedings of the 12th World Hydrogen Energy Conference*. pp. 669–678.

15. Hirano, A., Hon-Nami, K., Kunito, S., Hada, M., and Ogushi, T. (1998). Temperature Effect on Continuous Gasification of Microalgal Biomass: Theoretical Yield of Methanol Production and its Energy Balance, *Catalysis Today* 45. pp. 399–404.

16. Hofbauer, H., Rauch, R., Foscolo, P., and Matera, D. (2000). Hydrogen-rich Gas from Biomass Steam Gasification, *1st World Conference on Biomass for Energy and Industry,* Sevilla, Spain, June 2000. pp 1999–2001.

17. Iqbal, M., Dalai, A.K., Thring, R.W., and Bakhshi, N.N. (1998). Potential of Producing Hydrogen and High Btu Gas from Steam Gasification of Lignins, *33rd Intersociety Engineering Conference on Energy Conversion, Colorado Springs, CO, August 2–6* (IECEC-98-086).

18. Kubiak, H., Papamichalis, A., and van Heek, K.H. (1996). Production of Hydrogen by Allothermal Gasification of Biomass, *Hydrogen Energy Progress XI, Proceedings of the 11th World Hydrogen Energy Conference*; Vol. 1. pp. 621–629.

19. Lobachyov, K.V. and Richter, H.J. (1998). An Advanced Integrated Biomass Gasification and Molten Fuel Cell Power System, *Energy Convers. Manage. 39.* pp. 1931–1943.

20. McDonald, R.M., Swan, J.E., Donnelly, P.E., Mills, R.A., and Vaughan, S.R. (1981). Protein-extracted Grass and Lucerne as Feedstocks for Transport Fuel Production. *Transactions of the New Zealand Institution of Engineers, Electrical/Mechanical/Chemical Engineering,* 8, (2). pp. 59–64.

21. McKinley, K.R., Browne, S.H., Neill, D.R., Seki, A., and Takahashi, P.K. (1990). Hydrogen Fuel from Renewable Resources, *Energy Sources,* 12. pp. 105–110.

22. Milne, T.A., Abatzoglou, N., and Evans, R.J. (1998). Biomass Gasifier Tars: Their Nature, Formation and Conversion. November (NREL/TP-570-25357).

23. Naushkin, Ya. M., Zhorov, Yu. M., Nikanorva, L.P., Gorlov, E.G., and Vakhabov, O. (1988). Production of Hydrogen by Catalytic Conversion of Aqueous Suspensions of Plant Biomass with Steam, *Solid Fuel Chemistry (English Translation of Khimiya Tverdogo Topliva),* 22, (6). pp. 127–132.

24. Rapagna, S. and Foscolo, P.U. (1998). Catalytic Gasification of Biomass to Produce Hydrogen Rich Gas, *Int. J. Hydrogen Energy,* 23, (7). pp. 551–557.

25. Richardson, J.H., Rogers, R.S., Thorsness, C.B., and Wallman, P.H. (1995). Conversion of Municipal Solid Waste to Hydrogen, *Proceedings of the 1995 U.S. DOE Hydrogen Program Review,* Coral Gables, FL, April 18–21, 1995, Vol. 2 (NREL/CP-430-20036). pp. 731–755.

26. Rogers, R. III (1994). Hydrogen Production by Gasification of Municipal Solid Waste. Lawrence Livermore National Laboratory (UCRL-ID-117603).

27. Srinivas, S.T., Dalai, A.K., and Bakhshi, N.N. (1998). Potential of Producing Hydrogen and High Btu Gas from Steam Gasification of Biomass-derived Chars, *Canadian Society for Chemical Engineering Conference, London, Canada, 4–7 Oct*; CONF-981052(ISBN 0-920804-32-2).

28. Tsaros, C.L., Arora, J.L., and Burnham, K.B. (1976). The Manufacture of Hydrogen from Coal, *Proceedings of the 1st World Hydrogen Energy Conference,* Miami Beach, FL, March 1–3, 1976, Vol. 1. pp. 4A-3–4A-26.

29. Wallman, P.H., Richardson, J.H., Thorsness, C.B., Laninger, T.F., Klein, J.D., Winter, J.K., and Robin, A.M. (1996). Hydrogen Production from Municipal Solid Waste, *Proceedings of the 1996 U.S. DOE Hydrogen Program Review,* Miami, FL, May 1–3, 1996, NREL/CP-430-21968 (UCRL-JC-124574; CONF-9605195-2). pp. 481–498.

30. Wallman, P.H. and Thorsness, C.B. (1997). Hydrogen Production from Wastes, *Proceedings of the 1997 U.S. DOE Hydrogen Program Review,* Herndon, VA, May 21–23, 1997. pp. 165–176.

31. Shahbazov, Sh. S. and Usubov, I.M. (1996). Heliothermic Unit for Gasification Agricultural Product Wastes, *Hydrogen Energy Progress XI, Proceedings of the 11th World Hydrogen Energy Conference,* Vol. 1. pp. 951–954.

ENDNOTES

1. Klass, D.L., 1998. *Biomass for Renewable Energy, Fuels, and Chemicals.* Academic Press, San Diego, CA. 651 pp.

2. Quaak, P., H. Knoef, and H. Stassen, 1999. *Energy from Biomass. A Review of Combustion and Gasification Technologies.* World Bank Technical Paper No. 422, Energy Series, The World Bank, Washington, DC. 78 pp.

3. Karlschmitt, M. and A.V. Bridgwater, eds., 1997. *Biomass Gasification and Pyrolysis.* ISBM 1872691714. CPL Press, Newbury, UK. 550 pp.

4. Stassen, H.E.M., 1995. *Small-Scale Biomass Gasifiers for Heat and Power: A Global Review.* World Bank Technical Paper No. 296, World Bank, Washington, DC.

5. Miles, T.R., T.R. Miles Jr., L.L. Baxter, R.W. Bryers, B.M. Jenkens, and L.L. Oden, 1996b. *Alkali Deposits Found in Biomass Power Plants. Vol. II.* NREL/TP-433-8142. National Renewable Energy Laboratory, Golden, CO.

6. Jenkins, B.M., L.L. Baxter, T.R. Miles Jr., and T.R. Miles, 1998. *Combustion Properties of Biomass. Fuel Processing Technology, Vol. 54,* Elsevier Science, Amsterdam. pp. 17–46.

7. Baxter, L.L., T. Gale, S. Sinquefield, and G. Sclippa, 1997a. Influence of Ash Deposit Chemistry and Structure on Deposit Physical and Transport Properties, in *Developments in Thermochemical Biomass Conversion*, A.V. Bridgwater and D.G.B. Boocock, eds., Blackie Academic and Professional Press, London. pp. 1247–1262.

8. Baxter, L.L., T.R. Miles, T.R. Miles Jr., B.M. Jenkins, T.A. Milne, D. Dayton, R.W. Bryers, and L.L. Oden, 1997b. The Behavior of Inorganic Material in Biomass-Fired Boilers: An Overview of the Alkali Deposits Project, in *Developments in Thermochemical Biomass Conversion,* A.V. Bridgwater and D.G.B. Boocock, eds. Blackie Academic and Professional Press, London. pp. 1424–1446.

9. Baxter, L.L., 1992. *Ash Deposition During Biomass and Coal Combustion: A Mechanistic Approach.* SAND92-8216. Sandia National Laboratory, Albuquerque, NM.

10. Dayton, D., R. French, and T. Milne, 1994. Identification of Gas Phase Alkali Species Released During Biomass Combustion. *Proceedings of Bioenergy 94,* U.S. Department of Energy Western Regional Biomass Energy Program, Golden, CO. pp. 607–614.

11. Hauserman, W.B. and R.C. Timpe, 1992. Catalytic Gasification of Wood for Hydrogen and Methane Production. *Energy & Environmental Research Center,* University of North Dakota. pp. 1–38.

12. Cox, J.L., Tonkovich, A.Y., Elliott, D.C., Baker, E.G., and Hoffman, E.J., 1995. Hydrogen from Biomass: A Fresh Approach. *Proceedings of the Second Biomass Conference of the Americas* (NREL/CP-200-8098; CONF-9508104) August, Portland, OR. pp. 657–675.

13. Rapagna, S. and Foscolo, P.U., 1998. Catalytic Gasification of Biomass to Produce Hydrogen Rich Gas. *Int. J. Hydrogen Energy,* 23, (7). pp. 551–557.

14. Graham, R.G. and R. Bain, 1993. *Biomass Gasification: Hot Gas Clean-up.* Report Submitted to IEA Biomass Gasification Working Group, Ensyn Technologies/NREL, 44 pp.

15. Neeft, J.P.A., Knoef, H.A.M., and Onaji, P., 1999. *Behavior of Tars in Biomass Gasification Systems.* NOVEM EWAB Program Report 9919. Available from MHP Management Services, P.O. Box 127, 3950 AC Maarn, Netherlands. 75 pp.

4

Pyrolysis

INTRODUCTION

This chapter discusses the technology, practices, and emerging technologies of pyrolysis. Focus is given to the key applications of waste processing, and the application of coal and biomass pyrolysis processes for the production of synthetic fuels.

Pyrolysis (also known as carbonization, destructive distillation, dry distillation, or retorting) is the chemical decomposition induced in organic materials by heat in the absence of oxygen. In practice, it is not possible to achieve a completely oxygen-free atmosphere; pyrolytic systems are operated with less than stoichiometric quantities of oxygen. Because some oxygen will be present in any pyrolytic system, nominal oxidation does occur. If volatile or semivolatile materials are present, thermal desorption will also occur. Pyrolysis processes produce an array of solid and liquid derivatives, and fuel gases depending on the process conditions and the organic material content of the feedstock.

There are two major distinctions between combustion processes and pyrolysis. First, combustion processes are exothermic (i.e., they generate heat), whereas pyrolysis is an endothermic process (i.e., pyrolysis requires the addition of heat). Second, products of combustion processes are CO_2, water, and ash, whereas with pyrolysis the products are char, oil or tar, and fuel gases. These by-products can be used as fuel for combustion process or as intermediate products for synthesis and the production of chemicals.

Pyrolysis has the unique advantage of enabling the removal and separation of most impurities such as sulfur from fuels prior to their combustion.

Pyrolysis is neither new nor has its application been limited to small-scale production. It has traditionally been used in the manufacture of charcoal.[1] Prior to the 18th century, wood and charcoal fueled the smelters of the metallurgical industry.[2] In the 1700s, coal pyrolysis was practiced for the production of coke, as a replacement for charcoal in the metallurgical industry, and for the production of fuel gas for lighting. By the early 1800s, fuel gas produced by coal pyrolysis was used for street lighting, first in London and then New York, Baltimore, and other major cities of the world. Over time gasification and liquefaction technologies proved to be more economical for the conversion of coal to liquid and gases fuels.

The discovery of large reserves of low-cost petroleum ended large-scale commercial practices for the conversion of coal, oil shale, and tar sands to gases and liquid fuels by pyrolysis and other techniques. Despite this, development efforts for this technology have continued intermittently in the United States, Japan, Brazil, and a number of European Community countries in response to the world energy crisis and as a technology for waste disposal. Emphasis on environmental issues has renewed interest in low-temperature pyrolysis of biomass for the production of biofuels and high-temperature pyrolysis, including plasma pyrolysis, for the destruction of hazardous wastes.

---

[1] Klass, D.L, *Biomass for Renewable Energy, Fuels, and Chemicals,* Academic Press, 1998.

[2] Anderson, L.L. and Tillman, D.A., *Synthetic Fuels From Coal – Overview and Assessment,* Wiley Interscience, 1979.

TABLE 4.1 Biomass Pyrolysis Product Slate As A Function of Heating Rate, Residence Time, and Temperature

Technology	Heating Rate	Residence Time	Temperature °C	Products
Carbonation	Very low	Days	400	Charcoal
Conventional	Low	5–30 min	600	Oil, gas, char
Fast	Very high	0.5–5 s	650	Bio-oil
Flash-liquid	High	<1 s	<650	Bio-oil
Flash-gas	High	<1 s	<650	Chemicals, gas
Ultra	Very high	<0.5 s	1000	Chemicals, gas
Vacuum	Medium	2–30 s	400	Bio-oil
Hydro-pyrolysis	High	<10 s	<500	Bio-oil
Methanol-pyrolysis	High	<10 s	>700	Chemicals

Source: http://www.fao.org/docrep/T4470E/t4470e0a.htm

PYROLYSIS PRINCIPLES

General

Pyrolysis processes can be divided into two groups: low temperature and high temperature. The products of pyrolysis processes differ and can be controlled by temperature and the rate of material heating. Table 4.1 provides a summary of variations of the product slate for biomass and coal feedstocks.

At high temperatures the major product is gas, while at the low temperature the major product is tar or heavy oil. High heating rates minimize char formation by breaking down higher molecular species into gases products. Probstein and Hicks[3] provide a graphical summary of coal pyrolysis products as a function of heating rate, residence time, and temperature. The table shows that the gas and water yield is the highest for coal pyrolysis at high temperatures, when coal is heated to the desired temperature of 900 to 1000°C in a

---

[3] Probstein R.F. and Hicks R.E., *Synthetic Fuels,* McGraw-Hill Chemical Engineering Series, 1982, P. 106.

TABLE 4.2 Pyrolysis Product Yields in wt% From Different
Feedstocks at 500°C

Feedstock	Low Heating Value Gas	Oils	Ash/Char/Charcoal
Coal	56	31	10
Oil shale	4	16	80
Corncobs	17	22	26
Manure	20	18	28
MSW	23	11	<50
Paper	16	47	10
Wood chip	23	19	27

Note: The balance of the yield for each feedstock is water. Sources: Probstein and Hicks, 1982 and Klass 1988.

fraction of a second, and is the lowest when coal is heated to the same temperature over 1 to 2 hours.

When coal or biomass is heated, many reactions including dehydration, cracking, isomerization, dehydrogenation, aromatization, and condensations take place. Products are water, carbon dioxide, hydrogen, other gases, oils, tars, and char. The product yields vary, depending on the particular feedstock composition, particle size, heating rate, solids and gas residence times, and the reactor temperature.

As expected, the composition of feedstock can greatly impact the pyrolysis product yields. Table 4.2 reports product yields from pyrolysis of various biomass and fossil feedstocks at 500°C.

Experimental data with biomass show that devolatilization time increases with particle size.[4] The feedstock particle size affects the heating rate. Both heat flux and heating rates are lower in the larger particles than in the smaller ones.[5]

Large particles heat up more slowly; the average particle temperature is lower and therefore the devolatilization rate and yields are lower. Early work with coal particles show that

---

[4] See http://www.aidic.it/icheap6/webpapers/47%20Jand.pdf

[5] Zanzi, R., Sjöström, K. and Björnbom, E., **Rapid Pyrolysis of Agricultural** Residues at High Temperatures, submitted to Biomass & Bioenergy, 2001

TABLE **4.3**   Comparison of Slow and Rapid Pyrolysis Char Yield
and Density at 800°C to 1000°C

Feedstock	Slow Pyrolysis, Therombalance		Rapid Pyrolysis, Free Fall Reactor	
	Char Density (g/cm³)	Char Yield (wt% MAF)	Char Density (g/cm³)	Char Yield (wt% MAF)
Bagasse	n.d.	n.d.	n.d.	2.4
Olive waste	1.0		0.4	23.8
Pelletized straw	0.25	16.0	0.1	11.0
Wood (birch)		15.0	0.1	5.5

n.d.: not determined.
Source: http://www.lib.kth.se/Sammanfattningar/zanzi010607.pdf.

for small particles, less than 100 microns in size, particle size
is no longer a significant factor.[6]

Effect of Heating Rate

The effect of heating rate on pyrolysis yield is difficult to
measure while keeping other process conditions constant. It
is generally believed that rapid heating yields higher volatile
components than slower heating processes, as well as a more
reactive char. Experimental char yields and density for slow
and rapid pyrolysis of different biomass material presented
in Table 4.3 suggest a higher volatile yield and a less dense,
more porous, and more reactive char for rapid or fast pyrolysis
reactions. However, these data are affected by experimental
conditions.[7]

In slow heating experiments, slow removal of the volatiles
from the bed of biomass allows for secondary reactions between

---

[6] Elliott M.A., Editor, *Chemistry of Coal Utilization*, 2nd Supplementary
Volume, Wiley-Interscience, 1981, P.700.

[7] R. Zanzi, K. Sjostrom and Bjornbom, Royal Institute of Technology, Stockholm, Sweden – http://www.lib.kth.se/Sammanfattningar/zanzi010607.pdf

char particles and the volatiles promoting formation of second-ary char. The rapid removal of volatiles from the reactor in the rapid heating experiments prevents the secondary reactions from taking place, resulting in a lower char yield, i.e., higher volatile yield. The char is also less dense, more porous, and reactive because the secondary char does not form at active pour sites. Early experimental work by different researchers confirms that yield increases that were attributed to increased heating rates result primarily from different experimental conditions employed to achieve a higher heating rate.[8]

Effect of Temperature

Temperature significantly impacts pyrolysis product yield and composition. High temperatures promote gas production, while lower temperatures promote char and tar formation. The product slate from different biomass feedstock is similar at a given temperature, although the gas, liquid, and char yields can be very different depending on the feedstock (see Klass[9]). Literature data[10] indicate that gas yield and hydrogen content of the product gas increases with increasing temper-ature while the gas higher heating value (HHV) remains relatively constant. Increasing temperature also reduces char yield and volatile content.

Tables 4.4 and 4.5 show the effect of temperature on product yield and composition from the pyrolysis of mixed municipal solid waste (MSW) and bituminous coal. Other data[11] show that beyond a certain temperature char yield does

---

[8] Elliott M.A., Editor, *Chemistry of Coal Utilization*, 2nd Supplementary Volume, Wiley-Interscience, 1981, P.P. 694-695

[9] Klass, 1998; http://www.lib.kth.se/Sammanfattningar/zanzi010607.pdf

[10] Elliott M.A., Editor Chemistry of Coal Utilization, 2nd Supplementary Volume, 1981, P.P. 701-718

[11] Diebold J.P.,*Thermochemical Conversion of biomass to gasoline,* DOE's 11th Biomass Thermochemical Conversion Contractors Meeting, Richland, WA, Sept. 1980; http://www.lib.kth.se/Sammanfattningar/zanzi010607.pdf; Klass D. L and Emert G.E., *Fuels from Biomass and Wastes,* Ann Arbor Science, 1981.

TABLE 4.4 Sampling of Data Extracted From the Study By Klass (1998) Showing the Effects of Temperature on Product Yield and Gas Compositions From Pyrolysis of Combustible Fraction of MSW

	Temperature, °C		
	500	650	900
Product Yields and Recovery			
Gases, wt%	12.3	18.6	24.4
Liquids, wt%	61.1	59.2	58.7
Charcoal, wt%	24.7	21.8	17.7
Recovery, wt%	98.1	99.6	100.8
Gas Composition and HHV			
H_2, mol%	5.56	16.6	32.5
CO, mol%	33.5	30.5	35.3
CO_2, mol%	44.8	31.8	18.3
CH_2, mol%	12.4	13.9	10.5
C_2H_6, mol%	3.03	3.06	1.07
C_2H_6, mol%	0.45	2.18	2.43
HHV, MJ/m^3	12.3	15.8	15.1

not reduce for a given particle size, feedstock, and heat rate, indicating that a maximum gas yield can be obtained at a certain temperature unless the feedstock is heated to very high temperatures using plasma torch.

Plasma torches can heat the feedstocks to temperatures of 3000°C and higher, converting its organic content to gases and its inorganic content to a glassy slag. Plasma technology uses an electrical discharge "arc" to heat a gas to a very high temperature, usually above 3000°C. The heated gas is then used as a heat source for a particular application. In pyrolysis and gasification applications, the amount of oxygen in the plasma reactor must be carefully controlled. The extreme heat generated in a plasma reactor breaks down organic material to its basic components, primarily hydrogen, carbon monoxide, and carbon dioxide.

TABLE 4.5   Effects of Temperature On Product Yield and Gas
Compositions From Pyrolysis of Bituminous Coal

Parameters	Pyrolysis Temperature			
	400–440°C	620–690°C	740–780°C	1000°C
	Product Yields			
Gases, wt%	5.27	7.74	8.36	17.94
Liquids, wt%	1.18	9.45	15.0	25.4
Char, wt%	93.1	83.8	71.9	53.0
Recovery, wt%	99.6	101.0	95.4	97.3
	Gas Composition			
H_2, wt%	0.0	0.0	0.06	1.01
CO, wt%	0.0	0.21	0.40	2.42
CO_2, wt%	0.11	0.39	0.49	1.23
CH_4, wt%	0.0	0.22	0.59	2.49
CH_{gas}, wt%	0.0	0.90	1.56	3.95

Source: Elliott, M.A., Editor, *Chemistry of Coal Utilization*, 2nd Supplementary
Volume, Wiley-Interscience, 1981.

APPLICATIONS

Large-scale Commercial Processes for Mixed Solid Waste

Pyrolysis of coal and biomass (primarily wood) was practiced
commercially for the production of fuel-related gas and for
smokeless solid fuel in the manufacture of charcoal from the
1700s to early 1900s. Within the last decade pyrolysis processes have attracted renewed attention primarily due to
increased emphasis on pollution prevention and clean technologies. Environmental drivers include greenhouse gases
and the passage of more restrictive regulations for landfilling
of municipal and medical wastes in the European Community,
as well as more stringent air emission regulations. As discussed further on, pyrolysis has significant advantages over
incineration in terms of lower air emissions.

Mitsui Engineering and Shipbuilding Co., Ltd. (MES) of
Japan has developed and commercialized a pyrolysis process
for processing municipal solid waste (Figure 4.1). The process,

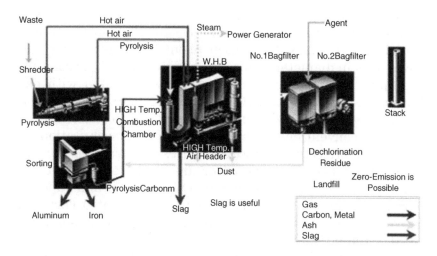

Figure 4.1   R21 process flow diagram. (*Source:* http://nett21.
unep.or.jp/JSIM_DATA/WASTE/WASTE_3/html/Doc_436.html)

known as Mitsui Recycling 21 (R21), utilizes a kiln to pyrolyze
waste material at 450°C.[12] Iron and aluminum are separated
from the solid residue, leaving the pyrolysis reactor, and then
the remaining solids are combusted along with the pyrolysis
gases at 1300°C in a slagging combustor in a low-oxygen
atmosphere (about 3.5%). Non-combustible material is
removed from the combustor as molten slag. A portion of hot
combustion gases are used to provide the heat input for pyrolysis reactions. The remaining heat is recovered to generate
steam for power generation.

In early 2002, two commercial plants with capacities of
70,000 and 150,000 ton/year were placed into operation in
Japan (*Power Magazine*, May/June 2002). Construction for
four plants ranging from about 58,000 ton/year to about
90,000 ton/year was also completed in Japan in late 2002 to
early 2003.

The largest operating pyrolysis facility in Europe is a
100,000 ton/year plant located northeast of Dortmund at

---

[12] see Web site www.mes.co.jp

Hamm, in Germany. The plant began operation in 2001 and is integrated with a conventional fossil-fueled power station, utilizing two rotary kilns each about 24 meters long and 2.8 meters in diameter. Kilns are charged with waste containing up to 50% plastics and are externally heated with gas burners to approximately 700°C in about 1 hour. The resulting pyrolysis gas and solid residue is then used to substitute up to 10% of conventional fossil fuel used in a power station.

Other projects for processing tires, medical wastes, and biomass are being considered or are at the early stages of development. For example, Fortum Oil & Gas Oy and Vapo Oy of Finland have constructed a pyrolysis pilot plant at Fortum's Provoo Refinery that produces 350 kg per hour of bio-oil from wood.[13] Liquid fuel production began in May 2002 at this plant, and the bio-oil is being used as heating fuel in large residential and public buildings and industry.

In the U.S., the largest pyrolysis plant is ENCOAL® Corp.'s 1000 ton coal/day pyrolysis plant, which began operation in 1992 under a cooperative agreement with the U.S. Department of Energy.[14]

Figure 4.2 presents a simplified flow diagram of the ENCOAL® Liquid from Coal (LFC) process. The process upgrades low-rank coals to two fuels, Process-Derived Coal (PDF™) and Coal-Derived Liquid (CDL™). Coal is first crushed and screened to about 50 mm by 3 mm and conveyed to a rotary grate dryer, where it is heated and dried by a hot gas stream under controlled conditions. The gas temperature and solids residence time are controlled so that the moisture content of the coal is reduced but pyrolysis reactions are not initiated. Under the drier operating conditions most of the coal moisture content is released; however, releases of methane, carbon dioxide, and monoxide are minimal. The dried coal is then transferred to a pyrolysis reactor, where hot recycled gas heats the coal to about 540°C. The solids residence time

---

[13] see http://www.vtt.fi/pro/pro2/pro22/pyrolysis.htm
[14] Coal Refineries, DOE, July, 1991; ENCOAL® Mild Coal Gasification Project, A DOE Assessment, NETL, March 2002

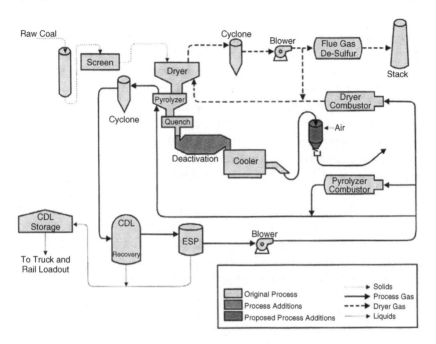

Figure 4.2 ENCOAL® Process Flow Diagram.

and the heating rate are controlled so that the remaining free water and volatile gaseous material are released. After leaving the pyrolysis reactor, solids are quenched to stop the reactions and are then further cooled in a vibrating fluidized bed and a rotary cooler. Drying gas stream temperature, oxygen content, and residence time are controlled to ensure deactivation of the coal within the vibrating fluidized-bed dryer. In the rotary cooler, a controlled amount of water is added to the solid products to rehydrate the PDF™ to near its equilibrium moisture content and to stabilize it.

After the pyrolysis gas stream passes through a cyclone and removes entrained particles, it is quenched in a tower to condense the final oil products or CDL™. The tower is designed to prevent water condensation; electrostatic precipitators (ESPs) recover any remaining liquid droplets and mist from the gas leaving the tower.

Some of the gas from the ESPs are burned in the pyro-lyzer combustor and blended with recycled gas to provide the heat for the pyrolyzer. A portion of the gas is burned in the dryer combustor and is then blended with a recycle stream to provide heating for the dryer. Any remaining gas leaving the ESPs is recycled to the pyrolyzer.

The gas stream leaving the dryer is treated in wet scrubbers, using a water-based sodium carbonate (Na_2CO_3) solution to recover fine particulates that escape the dryer cyclone and remove most of the SO_x from the flue gas.

The U.S. Department of Energy National Renewable Energy Laboratory (NREL) is also pursuing research focused on pyrolysis for production of bio-oil. However, there is no large-scale, commercial plant operating in the United Sates for production of bio-oil, although small-scale plants are reported to be operating intermittently in rural areas (NPR News Special Projects, i.e., http://npr.org/view_content/).

DynaMotive Energy Systems (Canada) has operated a 10 ton/day pyrolysis plant for the production of bio-oil from biomass since spring 2001. In May 2003, DynaMotive also announced plans to fabricate modular plants for converting 100 ton/day of dry biomass into 16,000 gallons of bio-oil in late 2003 and 2004; it also announced a 2.5 MW combined heat and power (CHP) plant in Ontario, Canada, utilizing a fast pyrolysis process developed by Resources Transforms International (RTI). Figure 4.3 shows a flow diagram of Dyna-Motive's process, known as BioTherm™. Biomass with less than 10% moisture and 1 to 2 mm particle size is fed into a bubbling fluidized-bed reactor. In the reactor, the biomass is heated to 450 to 500°C in the absence of oxygen, undergoes thermal and chemical reactions, and produces a gaseous stream and char. The gas is passed through a cyclone to remove char particles and then quenched to condense bio-oil. The process produces three products: bio-oil (60 to 75 wt%), char (15 to 25%), and non-condensable gases (10 to 20%). The non-condensable gases are recycled and supply a major part of the energy needed by the pyrolysis process. Bio-oil consists of 20 to 25% water, 25 to 30% water insoluble pyrolytic lignin, 5 to

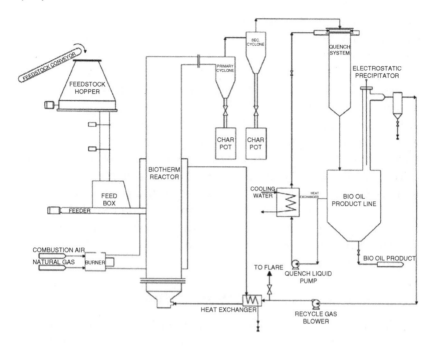

Figure 4.3 BioTherm™ Process Flow Diagram.

12% organic acids, 5 to 10% non-polar hydrocarbons, 5 to 10% anhydrosugars, and 5 to 25% other oxygenated compounds.[15]

Application to Contaminated Soil Remediation

Thermal processes like pyrolysis use heat to increase the volatility (separation); to burn, decompose, or detonate (destruction); or to melt (immobilization) contaminants in soil. Separation technologies include thermal desorption and hot gas decontamination. Destruction technologies include incineration, open burn/open detonation, and pyrolysis. Vitrification is used to immobilize inorganic compounds and to destroy some organic materials. In contrast, pyrolysis transforms

---

[15] BioTherm™ A System for Continuous Quality, Fast Pyrolysis BioOil, Dynamotive Energy Systems Corporation, 4th Biomass Conference of the Americas, Oakland, California, September 1, 1999

hazardous organic materials into gaseous components, small quantities of liquid, and a solid residue (coke) containing fixed carbon and ash.

Pyrolysis of organic materials produces combustible gases, including carbon monoxide, hydrogen and methane, and other hydrocarbons. If the off-gases are cooled, condensable gases, if any, produce an oil/tar residue and contaminated water. The pyrolysis gases require further treatment. The off-gases may be treated in a secondary combustion chamber, flared, and partially condensed. Particulate removal equipment such as fabric filters or wet scrubbers are also needed.

Conventional thermal treatment methods, such as rotary kiln, rotary hearth furnace, or fluidized-bed furnace, are used for waste pyrolysis. Molten salt process may also be used for waste pyrolysis.

Treatment of Municipal Solid Waste (MSW)

Currently, there are two MSW disposal facilities operating in Japan, and one is reported to be under construction in Rome using plasma technology. The first facility, located in Yoshii, Japan, was commissioned in 1999 and processes 24 ton/day of MSW in a single reactor. Developed by Hitachi Metal and Westinghouse Plasma Corp., it produces a syngas that is utilized in a waste-heat boiler to generate steam. The second facility, known as EcoValley Facility, is located in Utashini, Japan. It is similar to the Yoshii plant and was also developed by Hitachi Metal and Westinghouse Plasma but uses two 83 ton/day trains. It is designed to process automobile shredder residue and MSW. This plant was commissioned in late 2002. The third facility is being developed by ENEL, the major Italian utility, and Solena Group. Westinghouse Plasma is supplying the required plasma torches. The plant has a capacity of 336 ton/day and will generate electricity using Frame 6 combustion turbines from General Electric. The Italian government has guaranteed electricity prices at 14 cents per kWh above the prevailing rates due to the environmental benefits of the project.

Treatment of Medical Waste

The most widely used methods for medical waste processing are incineration and autoclaving. Problems with air pollution, among other disadvantages, have caused many government and state regulatory agencies to introduce more stringent air-quality standards. Healthcare and other facilities that generate medical waste have found that it is cost-prohibitive to meet the more stringent requirements through retrofitting existing incinerators or purchasing new equipment and have simply deactivated their incinerators.

A major concern associated with the use of autoclaves has been the generation of potentially hazardous chemicals. Since pressurized steam is an excellent method of volatilizing organic compounds, and many organic reactions are accelerated at elevated temperatures, a wide variety of organic compounds may be emitted depending upon the quantity and composition of the hazardous chemicals contained in the waste. This is an issue applicable to most medical waste treatment systems. Further, standard autoclaves cannot be used to treat a wide variety of waste (e.g., radioactive, chemotherapeutic waste categorized as hazardous waste) and pathologic waste. Recent advances in autoclave technology have created a few hybrid systems that have demonstrated their ability to treat pathological waste; however, these are limited commercially.

The reduction in the numbers of incinerators and the limitations of autoclaves have created the need for alternative medical waste treatment systems. Currently, there are over 40 such technologies available from greater than 70 manufacturers within the United States, Europe, the Middle East, and Australia. While these systems vary in their treatment capacity, the extent of automation, and overall volume reduction, all alternative technologies utilize one or more of the following methods: (1) heating the waste to a minimum of 90 to 95°C by means of microwaves, radio waves, hot oil, hot water, steam, or superheated gases; (2) exposing the waste to chemicals such as sodium hypochlorite (household bleach) or chlorine dioxide; (3) subjecting the waste to heated chemicals; and (4) exposing the medical waste to irradiation sources.

Thermal systems that use heat to inactivate pathogenic microorganisms are the most common alternative technologies for the treatment of medical waste. These systems can be broadly divided into those using low temperatures — 95°C (moist heat) to 250°C (dry heat) — and those that use high temperatures — from approximately 500°C to greater than 6000°C. The latter systems combust and destroy the waste as part of the treatment process.

Since pyrolysis involves the high-temperature (545 to 1000°C) treatment of the waste, systems treat, destroy, and reduce the volume of medical waste, e.g., Plasma Energy Pyrolysis System (PEPS). Plasma technology relies on an electric current to ionize an inert gas (e.g., argon), resulting in the formation of an electric arc to create temperatures as high as 6000°C. The medical waste within the system is brought to temperatures between 1300 to 1700°C, destroying potentially pathogenic microbes and converting the waste into a glassy rock or slag, ferrous metal, and inert gases, e.g., Peat's Plasma Arc Reactor System.

Plasma Torches and Plasma Pyrolysis

Plasma heating was originally developed for the metallurgy industry as an efficient alternative to conventional heating. During the early 1900s, plasma heaters were used in the chemical industry to manufacture acetylene fuel from natural gas. Today, plasma technology is being used successfully in industrial plants worldwide for different applications ranging from chemical and metallurgical industries to the waste/environment industry.

A plasma torch is a tubular device that converts electricity into heat via the resistance of a plasma. Plasma is a fourth state of matter: an ionized gas resulting, e.g., from electric discharges. The application of Plasma Pyrolysis with Vitrification (PPV) to municipal waste disposal causes the gasification and molecular dissociation of organic matter at the temperature of 2000°C. Adding some steam to the process causes gasification in a few milliseconds; there are no any intermediate reactions. Since there is no oxygen available there is no burning in any form, and no furans, dioxins, fumes, or ashes are formed. The result is a highly energetic synthesis gas, which is composed of 80% hydrogen and carbon monoxide.

The very high temperature (3000 to 7000°C) achievable with plasma arc torches makes this technology a viable tool for the thermal destruction of many types of wastes. As an example, when pure asbestos is subjected to temperatures above 1000°C, the asbestos fibers melt and subsequently solidify into a non-hazardous, chemically inert, solid material. The resulting slag-like substance is then suitable for disposal into regular landfills. Other types of waste will generate different product gas and slag characteristics. The chemical composition of different input waste materials will result in differing gas and slag characteristics. Input waste materials with a high carbon content and a high percentage of non-volatile material will produce results very similar to those with municipal solid waste. Other waste materials, such as biomass, liquid wastes, and organic wastes, will not produce much slag since virtually all of the waste is gasifiable. Table 4.6 provides a list of waste materials that can be treated with this technology.

Plasma gasification is a generic-type process that can accommodate virtually any input waste material in as-received condition, including liquids, gases, and solids in any form or combination. Also, moisture content is not a problem. Liquids, gases, and small particle-size waste materials are very easily and efficiently processed. Bulky items, such as household appliances, tires, and bedsprings, can also be readily accommodated without loss of destruction efficiency. The reactor vessel and waste feed mechanism are designed for the physical characteristics of the input waste stream. Even waste materials such as low-level radioactive waste can be processed to reduce the bulk and encapsulate the radioactive constituents to reduce leachability.

TABLE 4.6   Waste Materials That Can Be Treated by
Plasma Pyrolysis

Asbestos-containing material
Asbestos fibers
Automobile tires
Car fluff
Ceramic waste
Coal wastes
Contaminated landfill material
Contaminated soils and fines
Electric arc furnace dust
Ferro-manganese reduction
Ferrous chromium-containing waste
Glass waste
Harbor sludges
Hazardous fly ash
Hospital medical waste
Incinerator ash
Low-level radioactive waste
Mixed source waste (combination of different waste source
    with MSW, ash, coal, tires, etc.)
Municipal solid waste
Natural gas for acetylene production
Niobium recovery
Paints
Pathological wastes
PCB oil
Portland cement manufacture waste
Sludges
Solvents
Steel scrap
Titanium scrap melt
Waste coal

Plasma torch systems have the following components:

- A feed section, primary and secondary chambers, scrubber, and induced draft fan.
- Some system designs have a double port feeding mechanism that is charged with inert gas to prevent air venting into the hot zone.

- The plasma torch is usually mounted on a circular end wall. The torch is aligned in such a way that waste material falls into the hot zone of the plasma arc. The torch cathode and anode are cooled by forced water flow. In addition, auxiliary cooling of the anode with air can be incorporated in the plasma torch.

The pyrolysis product gas enters the secondary chamber through a burner that is mounted vertically, enabling flow of the product gas into the combustion chamber, ensuring extended flame length. The treated secondary off gas is scrubbed in a shower and ejected into the atmosphere.

A plasma torch is based on arc ignition between a thermionic tungsten cathode and a co-axial copper anode; both water-cooled anode and cathode are immersed in an axial magnetic field. Nitrogen is generally chosen as the plasma gas. Air or steam can be injected into the plasma to increase the enthalpy and to produce sub-stoichiometric incineration. The torch is powered by a thyristor-controlled rectifier, which has controls to match the torch impedance.

Magnetized non-transferred plasma torches have been adapted for pyrolysis technology. These torches are characterized by extended hot flame. Air plasma torches are also being developed to mitigate the logistic problems associated with nitrogen usage. The waste introduced into the extended hot zone of the primary chamber is pyrolyzed. The pyrolysis product gas contains CO and H_2 along with hydrocarbons and is combustible. This material is ignited in a burner inside the secondary chamber. The exhaust gas scrubbed and released into the atmosphere is almost colorless.

Since pyrolysis converts waste into CO, CH_4, and H_2, the product gases can be processed in an atmospheric pressure non-equilibrium plasma reformer to improve the energy-recovery potential of the product gas. Energy-recovery options include heat and chemical energy recovery.

Consistently low environmental emission characteristics exhibited by plasma gasification support the idea that it can be used as a waste treatment alternative to other technologies with significantly lower air emissions than competing technologies. Extremely tight physical and chemical bonds within the slag results in consistently low leachate characteristics.

ENDNOTES

1. Probstein, R.F. and Hicks, R.E., *Synthetic Fuels,* McGraw-Hill Chemical Engineering Series, 1982, p. 106.

2. See http://www.aidic.it/icheap6/webpapers/47%20Jand.pdf.

3. Zanzi, R., Sjöström, K., and Björnbom, E., Rapid Pyrolysis of Agricultural Residues at High Temperatures, submitted to *Biomass & Bioenergy,* 2001.

4. Elliott, M.A., Editor, *Chemistry of Coal Utilization*, 2nd Supplementary Volume, Wiley-Interscience, 1981, p. 700.

5. R. Zanzi, Sjostrom, K., and Bjornbom, Royal Institute of Technology, Stockholm, Sweden — http://www.lib.kth.se/Sammanfattningar/zanzi010607.pdf.

6. Elliott, M.A., Editor, *Chemistry of Coal Utilization*, 2nd Supplementary Volume, Wiley-Interscience, 1981, pp. 694–695.

7. Klass, 1998, http://www.lib.kth.se/Sammanfattningar/zanzi010 607.pdf.

8. Elliott, M.A., Editor, *Chemistry of Coal Utilization*, 2nd Supplementary Volume, Wiley-Interscience,1981, pp. 701–718.

9. Diebold J.P.,*Thermochemical Conversion of Biomass to Gasoline,* DOE's 11th Biomass Thermochemical Conversion Contractors Meeting, Richland, WA, Sept. 1980; http://www.lib.kth.se/Sammanfattningar/zanzi010607.pdf; Klass, D.L and Emert, G.E., *Fuels from Biomass and Wastes,* Ann Arbor Science, 1981.

10. See http://www.mes.co.jp.

11. See http://www.vtt.fi/pro/pro2/pro22/pyrolysis.htm.

12. Coal Refineries, DOE, July 1991; ENCOAL® Mild Coal Gasification Project, A DOE Assessment, NETL, March 2002.

13. BioTherm™ A System for Continuous Quality, Fast Pyrolysis BioOil, Dynamotive Energy Systems Corporation, 4th Biomass Conference of the Americas, Oakland, CA, September 1, 1999.

5

Gas Cleanup Technologies

INTRODUCTION

There are various technologies that can be used to remove unwanted components, including particulates, alkali, tars, sulfur, and ammonia, from the producer gas stream. The gas cleanup and conditioning technologies for gasification systems are outlined in this chapter.

In practice, the gasifier and the gas conditioning technologies must be considered as integrated systems.

OVERVIEW OF PARTICULATE REMOVAL TECHNOLOGIES

Particulate removal requirements vary depending on the use of the product gas. For example, particulate levels must be reduced to below 50 mg/Nm^3 for gas engines and below about 15 mg/Nm^3 (>5 μm) for turbines, and to perhaps 0.02 mg/Nm^3 for synthesis gas systems. The primary types of systems include cyclonic filters, barrier filters, electrostatic filters, and wet scrubbers.

Cyclonic filters are a primary means of removing bulk particulates from gas streams. They rely on centrifugal force to separate solids from the gas by directing the gas flow into a circular path. Because of inertia, the particulates are unable to follow the same path and are separated from the gas stream. Although the physics of particulate removal are complex, cyclone filters with predictable performance can be designed in a straightforward manner using theoretical and empirical techniques developed over many years.

Cyclonic filters (and closely related designs such as U-tubes) are employed as an initial gas cleanup step in most gasifier systems because they are effective and relatively inexpensive to operate. In circulating fluidized-bed or entrained-bed gasifiers, cyclones are an integral part of the reactor design, providing for separation of the bed material and other particulates from the gas stream.

Cyclone filters are effective at removing larger particles and can operate over a wide range of temperatures, limited primarily by the material of construction. Cyclone filters are often designed as multiple units in series (multi-clones). They can remove >90% of particulates above about 5 microns in diameter at minimal pressure drops of 0.01 atm. Partial removal of material in the 1.5 micron range is also possible, but cyclonic filters become ineffective with sub-micron particles.

Since cyclone filters can operate at elevated temperatures, the sensible heat in the product gas can be retained. Cyclone filters also remove condensed tars and alkali material from the gas stream, although the vaporized forms of those constituents remain in the gas stream. In practice, the separation of significant amounts of tars from the gas stream may be done sequentially by first removing particulates at higher temperatures, where tars remain vaporized. The gas stream is then cooled and condensed tars are removed. The step-wise approach reduces the tendency of particulate material to stick to tar-coated surfaces and contribute to plugging.

Cyclonic filters are used in many processes and are available commercially from many vendors.

Barrier filters include a range of porous materials that allow gases to penetrate but prevent the passage of particulates. These filters effectively remove small-diameter particulates in the range of 0.5 to 100 μm in diameter from gas streams. Barrier filters can be designed to remove almost any size of particulate, including those in the sub-micron range, but the pressure differential across the filter will increase as the pore size decreases.

As a result, there are technical and economic constraints that effectively limit particulate removal to about 0.5 μm in systems such as gasifiers that must handle large gas volumes.

Barrier filters are cleaned by periodically passing pulsing clean gas through the filter in the reverse direction of normal gas flow. To reduce the overall particulate load, these filters are typically placed downstream from cyclone filters. Barrier filters are effective for removing dry particulates but are less suitable for wet or sticky contaminants such as tars.

Tars have a tendency to cling to the filter surface and can undergo subsequent carbonization reactions that lead to fouling and plugging. Even in the absence of further decomposition, tars are difficult to remove from these materials. Examples of barrier filters suitable for biomass systems include rigid, porous-candle, or cross-flow filters constructed of metal or ceramic bag filters constructed of woven material, and packed-bed filters.

Rigid barrier filters (also called hot gas filters) provide an opportunity to produce a clean fuel gas while retaining the sensible heat of the fuel gas. In larger-scale gasification systems, these systems can potentially operate at moderate to high temperatures, depending on their material of construction, while removing small particulates. In most cases, cyclones would be used to reduce the overall loading of particulates before passing the gas through barrier filters for finishing.

Barrier filters are used in other industries for particulate removal, but they have been only recently tested in gasification systems. In large-scale systems, the use of metallic barrier filters may require partial product gas cooling to prevent sintering of the metal. In addition, metal filters are also susceptible to corrosion. Ceramic filters are suitable for higher temperature operation but are fragile and can break from thermal stress during temperature cycling. Ceramic filters are also susceptible to reactions with alkali vapors in gasification systems that can lead to decomposition or plugging.

Barrier filters have been tested and shown to have potential in gasification demonstration systems. Ceramic and metal candle-type filters were both tested at a commercial demonstration facility at Värnamo, Sweden, that is an integrated gasification combined cycle (IGCC) facility consisting of an air-blown, pressurized fluidized-bed gasifier coupled to an Ahlstrom gas turbine. The system capacity was approximately 6 MWe. In the tests at Värnamo, the gasifier gas was first cooled to about 350°C in a heat exchanger, and the resulting warm gas passed through candle filters. Some breakage of ceramic candle elements was encountered in early tests. These problems were caused in part because of the frequent thermal cycling in the demonstration facility that operated intermittently. Later tests at Värnamo were made with metallic candle filter elements, which are suitable for use with the warm gas at 350°C. The gasifier facility operated successfully in its full gasifier/gas turbine power generation mode for extended periods using these filters. The facility closed in 1999.

Bag filters are composed of woven material that intercepts small particles on the filter surface by impingement and electrostatic attraction. The efficiency of collection increases as the depth of the filter cake increases, making these filters highly efficient for small diameter, even sub-micron, particle size. Filters are periodically shaken or back-flushed to remove particulate accumulation. These filters are usually constructed of woven materials suitable for operation at low temperatures up to about 350°C.

These filters are well proven for removing particulates in a variety of systems but have not previously been used in biomass gasification systems. For gasification systems these filters require hot product gas to be cooled prior to filtration. For that reason, they are most appropriate for applications where retaining the sensible heat of the product gas is not critical. In addition, the presence of tars in the product gas can cause potential problems since tar condensation on the filter cake or the filter itself can lead to plugging. Care must be taken to ensure either that tars are removed from the stream prior to the bag filter or that the temperature remains high enough to prevent tar condensation.

Bag filters are being used or have been proposed for several medium- and large-scale gasifiers. These include systems using steam-based and gas turbine-based power generation technologies. In both types of systems, the product gases are typically cooled below 300°C and passed through bag filters.

Packed-bed filters may also act as barrier filters. In these systems, the raw product gases pass through a bed of packing material such as ceramic spheres or sawdust. Gas phase products pass through the bed, and particulates are de-entrained from the gas flow. Periodically, the particulates must be removed from the filter bed, typically by back-flushing. Small-scale gasifier facilities have used packed beds of materials such as sawdust or activated charcoal to remove both particulates and tars from the gas stream. In larger systems, the problem with accumulation of particulates or tars in the packed bed present potential operability problems. As a result, these types of filters are not being actively incorporated into larger-scale gasifier systems at this time.

Electrostatic filters have also been used extensively in a variety of gas cleaning operations. In these systems, the product gas flows past high-voltage electrodes that impart an electric charge to particulates, but do not affect the permanent gases. The particulates are then collected as the gas stream passes collector plates of the opposite polarity. The electrically charged particulates migrate to the collector plate and deposit on the surface. Particulates are removed from the scrubber plates by either wet or dry methods. Dry scrubbers use mechanical action to periodically remove material from the surface and can operate at temperatures of 500°C or more. Wet scrubbers remove particulates with a thin film of flowing water and are limited to temperatures of about 65°C.

Electrostatic filters have been used in many coal-fired power stations, and they have been used in some biomass combustion facilities. Their use in medium- or large-scale gasification systems is limited. Electrostatic filters are best suited for large-scale operation due to their physical size and cost, and the primary impediment to their use in current gasification systems is an economic one.

Wet scrubbers use liquid sprays, usually water, to remove particulates. Particles are collected by collision with liquid droplets, and the droplets are then de-entrained from the gas stream in a demister. The most common wet scrubbers use a Venturi design to create a pressure drop that allows solutions to be easily sprayed into the gas stream. Gas velocities are typically 60 to 125 m/sec in this throat area. Particulate removal efficiency is proportional to the pressure drop across the Venturi. With pressure drops of 0.25–2.5 kPa, these scrubbers can remove 99.9% of particles over 2 μm, and 95 to 99% of those over 1 μm. The wet particulates from the Venturi are subsequently removed from the gas stream by a demister. The demister can be a cyclone, packed-bed, or other type of collector.

Wet scrubbing requires that the water remain in the liquid phase, which requires that the product gas be cooled to below 100°C. This loss of sensible heat may be undesirable in some systems. Most gasification systems that currently use wet scrubbers do so primarily as a means to remove tars rather than particulates from the gas stream. Removing the particulates separately can prevent condensation of the sticky tars on the particulate surface, and that can prevent fouling and plugging of filter surfaces.

The following subsections provide the reader with details on control systems suitable for gasification. Selection of proper equipment depends on the gasification process, the nature and composition of the producer gas, the intended application of the producer gas, and the subsequent requirements for downstream applications.

PARTICULATE COLLECTION TECHNOLOGIES

Settling chambers, low-pressure-drop cyclones, and dynamic precipitators are among the most common and important devices applied in end-of-pipe control applications for dust and particulate matter.

Gravity Settling Chambers

Gravity settling chambers are the oldest and simplest means of removing suspended particles from a gas. In principle, pollutants are removed by reducing the velocity of the gas stream sufficiently to allow particles to settle out under the

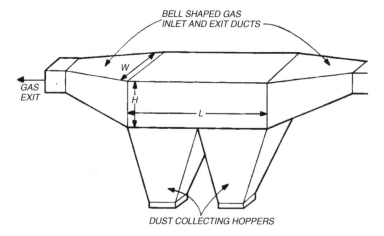

Figure 5.1   Gravity settling chamber features.

influence of gravity. The simplest chamber is merely a horizontal duct in which large particles settle out on the floor. High-efficiency settling chambers are often fitted with baffles or deflectors to change the gas flow direction.

This type of technology is a part of the group of air pollution controls collectively referred to as "precleaners," because they are oftentimes used to reduce the inlet loading of particulate matter (PM) to downstream collection devices by removing larger, abrasive particles. Settling chambers are also referred to as gravity settling chambers, gravity collectors, expansion chambers, and outfall chambers. Multiple-tray settling chambers are also referred to as Howard settling chambers.

Before selecting or sizing a specific control device, a careful evaluation of all aspects of the process and contaminants must be made. Improper terminology can lead to poor design or operation of any type of device.

Figure 5.1 is a simplified representation of a horizontal simple gravity settling chamber. It is a long duct fitted with hoppers on the floor to collect particulates. Physical dimensions are characterized by the ductwork above the collection hoppers: length = L, width = W, and height = H.

The primary section of the chamber is characterized by its cross-sectional area (W * H) and by its length (L). The cross-sectional area is designed to be larger than the inlet and exit ducts in order to reduce substantially the gas stream's inlet linear velocity. The length of the chamber determines the amount of time the particles remain at the reduced rate. This starving of the gas's forward motion allows the particles sufficient time to settle out into the hoppers.

For dilute systems, Stoke's law is applicable to particle settling. References cited at the end of this chapter provide design and sizing information.

In specifying settling chamber dimensions, gas flow velocities must be maintained below the re-entrainment velocity (pick-up velocity) of deposited particulate. As a general guideline, linear gas velocities are kept below 10 ft/sec (600 ft/min = 304.8 cm/sec). This is satisfactory for most materials; however, some low-density particulates are re-entrained at lower velocities.

Settling efficiency is normally related through a theoretical efficiency curve for a chamber of the dimensions sized. An example is illustrated in Figure 5.2.

There may be additional factors other than space availability that may enter into sizing considerations. For example, economics may dictate the use of only a gravity settling chamber for pollution control purpose. In such a case the removal of a wider range of particle sizes may be required. Rapid removal of larger particles could be accomplished by increasing the width of the chamber, and removal of the smaller-sized particles could be accomplished by increasing the length. Several efficiency curves can be generated for various chamber dimensions, and an optimum design based on legal aspects and economics can be made.

As noted, settling chambers are used to control PM, and primarily PM greater than 10 micrometers (μm) in aerodynamic diameter. Most designs only effectively collect PM greater than approximately 50 μm (see Wark[1]).

---

[1] Wark, 1981. Kenneth Wark and Cecil Warner, *Air Pollution: Its Origin and Control*, Harper Collins, New York, 1981.

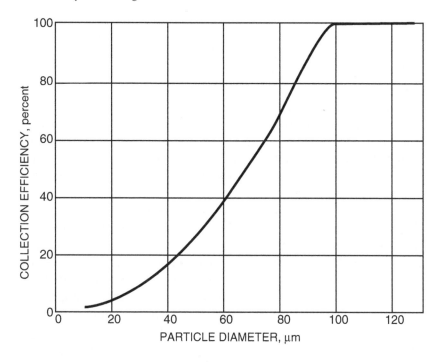

Figure 5.2 Generalized collection efficiency plot.

The collection efficiency of settling chambers varies as a function of particle size and settling chamber design. Settling chambers are most effective for large and/or dense particles. Gravitational force may be employed to remove particles where the settling velocity is greater than about 13 cm/s (25 ft/min). In general, this applies to particles larger than 50 μm if the particle density is low, down to 10 μm if the material density is reasonably high. Particles smaller than this would require excessive horizontal flow distances, which would lead to excessive chamber volumes. The collection efficiency for PM less than or equal to 10 μm in aerodynamic diameter (PM_{10}) is typically less than 10%. Multiple-tray chambers have lower volume requirements for the collection of particles as small as 15 μm.

Despite low collection efficiencies, settling chambers have been used extensively in the past in many conventional air pollution control applications. The metals refining industries have used settling chambers to collect large particles, such as arsenic trioxide from the smelting of copper ore. Power and heating plants have used settling chambers to collect large unburned carbon particles for reinjection into the boiler. They are particularly useful for industries that also need to cool the gas stream prior to treatment in a fabric filter.

Settling chambers have been used to prevent excessive abrasion and dust loading in primary collection devices by removing large particles from the gas stream, such as either very high dust loadings or extremely coarse particles that might damage a downstream collector in series with the settling chamber. The upstream use of settling chambers has declined with improvements in acceptable loading of other, more efficient control devices and increasing space restrictions at facilities. In cases where sparks or heated material is present in the waste gas, settling chambers are still used to serve as "spark traps" to prevent a downstream baghouse or filter from catching fire. These devices are generally constructed for a specific application from duct materials, though almost any material can be used. Settling chambers have been replaced, for most applications, by cyclones primarily due to the lower space requirements and the higher collection efficiency of cyclones. Multiple-tray settling chambers have never been widely used because of the difficulty in removing the settled dust from the horizontal trays.

The simple design and construction of settling chambers allows for almost any size and waste gas flow rate, but size is usually restricted to a 4.25 m² (14 ft²) shipping size. Units restricted by this shipping constraint will generally have flow rates that range up to 50 standard cubic meters per second (scm³/sec) (106,000 standard cubic feet per minute scfm). Typical settling chamber waste gas flow capacity is 0.25 to 0.5 sm³/sec per cubic meter of chamber volume (15 to 30 scfm per cubic foot of chamber volume).

Inlet gas temperatures are only limited by the materials of construction of the settling chamber and have been operated at temperatures as high as 540°C (1000°F).

Gas pollutant loadings can range from 20 to 4500 grams per standard cubic meter (g/sm³) (9 to 1970 grains per standard cubic foot (gr/scf)). Multiple-tray settling chambers can only handle inlet dust concentrations of less than approximately 2.3 g/sm³ (1.0 gr/scf).

Leakage of cold air into a settling chamber can cause local gas quenching and condensation. Condensation can cause corrosion, dust buildup, and plugging of the hopper or dust removal system. The use of thermal insulation can reduce heat loss and prevent condensation by maintaining the internal device temperature of the above the dew point.[2] (EPA, 1982). No pretreatment is necessary for settling chambers.

The following are cost ranges (expressed in third-quarter 1995 dollars) for a single conventional expansion-type settling chamber under typical operating conditions, developed using a modified EPA cost-estimating spreadsheet and referenced to the volumetric flow rate of the waste stream treated. For purposes of calculating the example cost effectiveness, flow rates are assumed to be between 0.25 and 50 sm³/sec (530 and 106,000 scfm), the inlet PM loading concentration is assumed to range from approximately 20 to 4500 g/sm³ (9 to 1970 gr/scf), and the control efficiency is assumed to be 50%. The costs do not include costs for disposal or transport of collected material. Capital costs can be higher than in the ranges shown for applications that require expensive materials. As a rule, smaller units controlling a low concentration waste stream will be more expensive (per unit volumetric flow rate) than a large unit cleaning a high pollutant load flow.

- Capital Cost: $330 to $10,900 per sm³/sec ($0.16 to $5.10 per scfm)
- Operational and Maintenance (O&M) Cost: $13 to $470 per sm³/sec ($0.01 to $0.22 per scfm), annually
- Annualized Cost. $40 to $1,350 per sm³/sec ($0.02 to $0.64 per scfm), annually
- Cost Effectiveness: $0.01 to $3.90 per metric ton ($0.01 to $3.50 per short ton), annualized cost per ton per year of pollutant controlled

---

[2] EPA, 1982.

As noted there are two primary types of settling chambers: the expansion chamber and the multipletray chamber. In the expansion chamber, the velocity of the gas stream is significantly reduced as the gas expands in a large chamber. The reduction in velocity allows larger particles to settle out of the gas stream. A multiple-tray settling chamber is an expansion chamber with a number of thin trays closely spaced within the chamber, which causes the gas to flow horizontally between them. While the gas velocity is increased slightly in a multiple-tray chamber, when compared to a simile extension chamber, the collection efficiency generally improves because the particles have a much shorter distance to fall before they are collected. Multiple-tray settling chambers have lower volume requirements than expansion-type settling chambers for the collection of small particles (~ 15 μm).

The efficiency of settling chambers increases with residence time of the waste gas in the chamber. Because of this, settling chambers are often operated at the lowest possible gas velocities. In reality, the gas velocity must be low enough to prevent dust from becoming re-entrained, but not so low that the chamber becomes unreasonably large. The size of the unit is generally driven by the desired gas velocity within the unit, which should be less than 3 m/s (10 ft/sec), and preferably less than 0.3 m/s (1 ft/sec).

Advantages of settling chambers include:

- Low capital cost
- Very low energy cost
- No moving parts; therefore, few maintenance requirements and low operating costs
- Excellent reliability
- Low pressure drop through device
- Device not subject to abrasion due to low gas velocity
- Provide incidental cooling of gas stream
- Temperature and pressure limitations are only dependent on the materials of construction
- Dry collection and disposal

Disadvantages of settling chambers include:

- Relatively low PM collection efficiencies, particularly for PM less than 50 µm in size
- Unable to handle sticky or tacky materials
- Large physical size
- Trays in multiple-tray settling chamber may warp during high-temperature operations

The most common failure mode of settling chambers is plugging of the chamber with collected dust. In expansion settling chambers the plugging can result from hopper bridging or hopper discharge seal failure. Multiple-tray settling chambers may experience plugging of the individual gas passages. Such failures can be prevented or minimized by use of hopper level indicators or by continuous monitoring of the dust discharge. Scheduled internal inspection can determine areas of air leakage and condensation, both of which may cause hopper bridging. Nominal instrumentation for a settling chamber generally includes only an indicator of differential static pressure. An increase in static pressure drop can indicate plugging.

Cyclone Separators

This type of technology is a part of the group of controls collectively referred to as "precleaners," because they are oftentimes used to reduce the inlet loading of particulate matter to downstream collection devices by removing larger, abrasive particles. Cyclones are also referred to as cyclone collectors, cyclone separators, centrifugal separators, and inertial separators. In applications where many small cyclones are operating in parallel, the entire system is called a multiple tube cyclone, multicyclone, or multiclone. Key features of the equipment are illustrated in Figures 5.3, 5.4, and 5.5.

Particle removal from a gas stream is achieved by centrifugal and inertial forces, induced by forcing particulate-laden gas to change direction. Cyclones are used to control PM, and primarily PM greater than 10 µm in aerodynamic diameter. However, there are high-efficiency cyclones designed to be effective for PM less than or equal to 10 µm

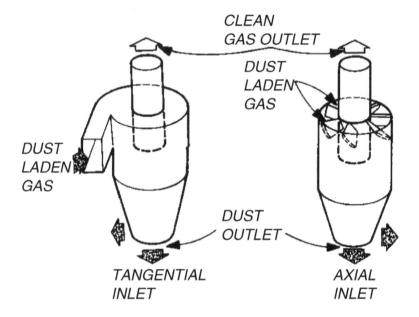

Figure 5.3 Configuration and inlet schemes.

and less than or equal to 2.5 μm in aerodynamic diameter (PM_{10} and $PM_{2.5}$). Although cyclones may be used to collect particles larger than 200 μm, gravity settling chambers or simple momentum separators are usually satisfactory and less subject to abrasion.

The collection efficiency of cyclones varies as a function of particle size and cyclone design. Cyclone efficiency generally increases with (1) particle size or density, (2) inlet duct velocity, (3) cyclone body length, (4) number of gas revolutions in the cyclone, (5) ratio of cyclone body diameter to gas exit diameter, (6) dust loading, and (7) smoothness of the cyclone inner wall. Cyclone efficiency will decrease with increases in (1) gas viscosity, (2) body diameter, (3) gas exit diameter, (4) gas inlet duct area, and (5) gas density. A common factor contributing to decreased control efficiencies in cyclones is leakage of air into the dust outlet.

Control efficiency ranges for single cyclones are often based on three classifications of cyclone; i.e., conventional,

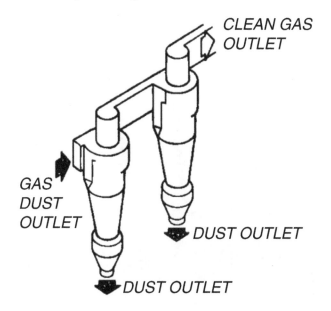

Figure 5.4 Series configuration.

high-efficiency, and high-throughput. The control efficiency range for conventional single cyclones is estimated to be 70 to 90% for PM, 30 to 90% for PM_{10}, and 0 to 40% for $PM_{2.5}$.

High-efficiency single cyclones are designed to achieve higher control of smaller particles than conventional cyclones. High-efficiency single cyclones can remove 5 μm particles at up to 90% efficiency, with higher efficiencies achievable for larger particles. The control efficiency ranges for high-efficiency single cyclones are 80 to 99% for PM, 60 to 95% for PM_{10}, and 20 to 70% for $PM_{2.5}$. Higher efficiency cyclones come with higher pressure drops, which require higher energy costs to move the waste gas through the cyclone. Cyclone design is generally driven by a specified pressure-drop limitation, rather than by meeting a specified control efficiency.

High-throughput cyclones are only guaranteed to remove particles greater than 20 μm, although collection of smaller particles does occur to some extent. The control efficiency ranges for high-throughput cyclones are 80 to 99% for PM, 10 to 40% for PM_{10}, and 0 to 10% for $PM_{2.5}$. Multicyclones are

Figure 5.5   Inlet manifold to multiclone.

reported to achieve from 80 to 95% collection efficiency for 5-μm particles.

Cyclones have been employed in numerous conventional air pollution and product recovery applications. Cyclones themselves are generally not adequate to meet stringent air pollution regulations, but they serve an important purpose as precleaners for more expensive final control devices such as fabric filters or electrostatic precipitators (ESPs). Cyclones are used extensively after spray-drying operations in the food and chemical industries, and after crushing, grinding, and calcining operations in the mineral and chemical industries to collect salable or useful material. In the ferrous and non-ferrous metallurgical industries, cyclones are often used as a first stage in the control of PM emissions from sinter plants, roasters, kilns, and furnaces. PM from the fluid-cracking process is removed by cyclones to facilitate catalyst recycling. Fossil-fuel and wood-waste fired industrial and commercial fuel combustion units commonly use multiple cyclones (generally upstream of a wet scrubber, ESP, or fabric filter) that collect fine PM (<2.5 μm) with greater efficiency than a single cyclone. In some cases, collected fly ash is reinjected into the combustion unit to improve PM control efficiency.

Typical gas flow rates for a single cyclone unit are 0.5 to 12 standard cubic meters per second (sm^3/sec) (1,060 to 25,400 scfm). Flows at the high end of this range and higher (up to approximately 50 sm^3/sec or 106,000 scfm) use multiple cyclones in parallel. There are single cyclone units employed for specialized applications that have flow rates of up to approximately 30 sm^3/sec (63,500 scfm) and as low as 0.0005 sm^3/sec (1.1 scfm).

Inlet gas temperatures are only limited by the materials of construction of the cyclone and have been operated at temperatures as high as 540°C (1000°F).

Gas loadings typically range from 2.3 to 230 grams per standard cubic meter (g/sm^3) (1.0 to 100 gr/scf). For specialized applications, loadings can be as high as 16,000 g/sm^3 (7,000 gr/scf), and as low as 1 g/sm^3 (0.44 gr/scf).

Cyclones perform more efficiently with higher particulate loadings, provided that the device does not become choked. Higher pollutant loadings are generally associated with higher flow designs.

The following are cost ranges (expressed in third-quarter 1995 dollars) for a single conventional cyclone under typical operating conditions, developed using an EPA cost-estimating spreadsheet,[3] and referenced to the volumetric flow rate of the waste stream treated. For purposes of calculating the example cost effectiveness, flow rates are assumed to be between 0.5 and 12 sm^3/sec (1,060 and 25,400 scfm), the PM inlet loading is assumed to be approximately 2.3 and 230 g/sm^3 (1.0 to 100 gr/scf), and the control efficiency is assumed to be 90%. The costs do not include costs for disposal or transport of collected material. Capital costs can be higher than in the ranges shown for applications that require expensive materials.

As a rule, smaller units controlling a waste stream with a low PM concentration will be more expensive (per unit volumetric flow rate and per quantity of pollutant controlled) than a large unit controlling a waste stream with a high PM concentration.

- Capital Cost: $4,200 to $5,100 per sm^3/sec ($2.00 to $2.40 per scfm)
- O&M Cost: $2,400 to $27,800 per sm^3/sec ($1.10 to $13.10 per scfm), annually
- Annualized Cost: $2,800 to $28,300 per sm^3/sec ($1.30 to $13.40 per scfm), annually
- Cost Effectiveness: $0.45 to $460 per metric ton ($0.41 to $420 per short ton), annualized cost per ton per year of pollutant controlled

Flow rates higher than approximately 10 sm^3/sec (21,200 scfm), and up to approximately 50 sm^3/sec (106,000 scfm), usually employ multiple cyclones operating in parallel. Assuming the same range of pollutant loading and an efficiency of 90%, the following cost ranges (expressed in third-quarter

---

[3] EPA, 1996.

1995 dollars) were developed for multiple cyclones, using an EPA cost-estimating spreadsheet,[4] and referenced to the volumetric flow rate of the waste stream treated.

- Capital Cost: $4,100 to $5,000 per sm³/sec ($2.00 to $2.40 per scfm)
- O&M Cost: $1,600 to $2,600 per sm³/sec ($0.75 to $1.20 per scfm), annually
- Annualized Cost: $2,000 to $3,100 per sm³/sec ($0.90 to $1.50 per scfm), annually
- Cost Effectiveness: $0.32 to $50 per metric ton ($0.29 to $46 per short ton), annualized cost per ton per year of pollutant controlled

Cyclones use inertia to remove particles from the gas stream. The cyclone imparts centrifugal force on the gas stream, usually within a conical shaped chamber. Cyclones operate by creating a double vortex inside the cyclone body. The incoming gas is forced into circular motion down the cyclone near the inner surface of the cyclone tube. At the bottom of the cyclone, the gas turns and spirals up through the center of the tube and out of the top of the cyclone. Particles in the gas stream are forced toward the cyclone walls by the centrifugal force of the spinning gas but are opposed by the fluid drag force of the gas traveling through and out of the cyclone. For large particles, inertial momentum overcomes the fluid drag force so that the particles reach the cyclone walls and are collected. For small particles, the fluid drag force overwhelms the inertial momentum and causes these particles to leave the cyclone with the exiting gas. Gravity also causes the larger particles that reach the cyclone walls to travel down into a bottom hopper. While they rely on the same separation mechanism as momentum separators, cyclones are more effective because they have a more complex gas flow pattern. Cyclones are generally classified into four types, depending on how the gas stream is introduced into the device and how the collected dust is discharged. The four

---

[4] EPA, 1996.

types are tangential inlet, axial discharge; axial inlet, axial discharge; tangential inlet, peripheral discharge; and axial inlet, peripheral discharge. The first two types are the most common. Pressure drop is an important parameter because it relates directly to operating costs and control efficiency. Higher control efficiencies for a given cyclone can be obtained by higher inlet velocities, but this also increases the pressure drop. In general, 18.3 meters per second (60 feet per second) is considered the best operating velocity. Common ranges of pressure drops for cyclones are 0.5 to 1 kPa (2 to 4 in. H_2O) for low-efficiency units (high throughput), 1 to 1.5 kPa (4 to 6 in. H_2O) for medium-efficiency units (conventional), and 2 to 2.5 kPa (8 to 10 in. H_2O) for high-efficiency units.

When high efficiency (which requires small cyclone diameter) and large throughput are both desired, a number of cyclones can be operated in parallel. In a multiple-tube cyclone, the housing contains a large number of tubes that have a common gas inlet and outlet in the chamber. The gas enters the tubes through axial inlet vanes, which impart a circular motion.

Another high-efficiency unit, the wet cyclonic separator, uses centrifugal force to enhance control efficiency.

Advantages of cyclones include:

- Low capital cost
- No moving parts; therefore, few maintenance requirements and low operating costs
- Relatively low pressure drop (2 to 6 inches water column), compared to amount of PM removed
- Temperature and pressure limitations are only dependent on the materials of construction
- Dry collection and disposal
- Relatively small space requirements

Disadvantages of cyclones include:

- Relatively low PM collection efficiencies, particularly for PM less than 10 µm in size
- Unable to handle sticky or tacky materials
- High-efficiency units may experience high pressure drops

Using multiple cyclones, either in parallel or in series, to treat a large volume of gas results in higher efficiencies, but at the cost of a significant increase in pressure drop. Higher pressure drops translate to higher energy usage and operating costs. Several designs should be considered to achieve the optimum combination of collection efficiency and pressure drop.

Fabric Filter Pulse Jet-Cleaned Type

This equipment is used for the capture of particulate matter, including particulate matter less than or equal to 10 μm in aerodynamic diameter (PM_{10}) and particulate matter less than or equal to 2.5 μm in aerodynamic diameter ($PM_{2.5}$). Typical new equipment design efficiencies are between 99 and 99.9%. Older existing equipment has a range of actual operating efficiencies of 95 to 99.9%. Several factors determine fabric filter collection efficiency. These include gas filtration velocity, particle characteristics, fabric characteristics, and cleaning mechanism. In general, collection efficiency increases with increasing filtration velocity and particle size. For a given combination of filter design and dust, the effluent particle concentration from a fabric filter is nearly constant, whereas the overall efficiency is more likely to vary with particulate loading. For this reason, fabric filters can be considered to be constant outlet devices rather than constant efficiency devices. Constant effluent concentration is achieved because at any given time, part of the fabric filter is being cleaned. As a result of the cleaning mechanisms used in fabric filters, the collection efficiency is constantly changing. Each cleaning cycle removes at least some of the filter cake and loosens particles that remain on the filter. When filtration resumes, the filtering capability has been reduced because of the lost filter cake and loose particles are pushed through the filter by the flow of gas. As particles are captured, the efficiency increases until the next cleaning cycle. Average collection efficiencies for fabric filters are usually determined from tests that cover a number of cleaning cycles at a constant inlet loading. Note that this equipment is applicable to stationary point source control.

Fabric filters have had a long history in performing effectively in a wide variety of conventional applications. Common applications using pulse-jet cleaning systems include:

- Utility boilers (coal)
- Industrial boilers (coal, wood)
- Commercial/Institutional boilers (coal, wood)
- Ferrous metal processing (e.g., iron and steel production and steel foundries)
- Mineral products (e.g., cement manufacturing, coal cleaning, and stone quarrying and processing)
- Asphalt manufacturing
- Grain milling

Fabric filters come in many different sizes and configurations. In older plants one can find makeshift operations. Refer to Figures 5.6 and 5.7 for examples of common configurations and characteristics.

Baghouses are separated into two groups, standard and custom, which are separated into low, medium, and high capacity. Standard baghouses are factory-built, off-the-shelf units. They may handle from less than 0.10 to more than 50 sm^3/sec ("hundreds" to more than 100,000 scfm). Custom baghouses are designed for specific applications and are built to the specifications prescribed by the customer. These units are generally much larger than standard units, i.e., from 50 to over 500 sm^3/sec (100,000 to over 1,000,000 scfm).

Typically, gas temperatures up to about 260°C (500°F), with surges to about 290°C (550°F), can be accommodated routinely, with the appropriate fabric material. Spray coolers or dilution air can be used to lower the temperature of the pollutant stream. This prevents the temperature limits of the fabric from being exceeded. Lowering the temperature, however, increases the humidity of the pollutant stream. Therefore, the minimum temperature of the pollutant stream must remain above the dew point of any condensable in the stream. The baghouse and associated ductwork should be insulated and possibly heated if condensation may occur.

Typical inlet concentrations to baghouses are 1 to 23 g/m^3 (0.5 to 10 gr/ft^3), but in extreme cases, inlet conditions may

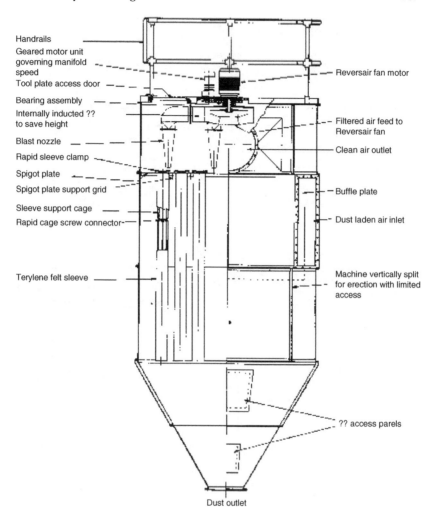

Handrails

Geared motor unit governing manifold speed

Tool plate access door

Bearing assembly

Internally inducted ?? to save height

Blast nozzle

Rapid sleeve clamp

Spigot plate

Spigot plate support grid

Sleeve support cage

Rapid cage screw connector

Terylene felt sleeve

Reversair fan motor

Filtered air feed to Reversair fan

Clean air outlet

Buffle plate

Dust laden air inlet

Machine vertically split for erection with limited access

?? access parels

Dust outlet

Figure 5.6 European design.

vary between 0.1 to more than 230 g/m^3 (0.05 to more than 100 gr/ft^3).

Moisture and corrosives content are the major gas stream characteristics requiring design consideration. Standard fabric filters can be used in pressure or vacuum service, but only within the range of about ± 640 mm of water column (25 inches of water column). Well-designed and operated baghouses have

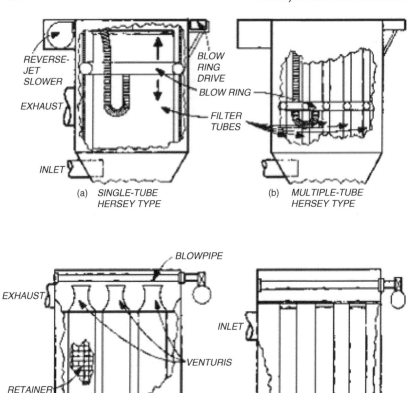

Figure 5.7 Pulse jet collector type.

been shown to be capable of reducing overall particulate emissions to less than 0.05 g/m^3 (0.010 gr/ft^3), and in a number of cases, to as low as 0.002 to 0.011 g/m^3 (0.001 to 0.0115 gr/ft^3).

Cost estimates are presented below for pulse jet-cleaned fabric filters. The costs are expressed in fourth-quarter 1998 dollars. The cost estimates assume a conventional design under typical operating conditions and do not include auxiliary equipment such as fans and ductwork. The costs for pulse jet-cleaned systems are generated using EPA's cost-estimating spreadsheet for fabric filters. Costs are primarily driven by the waste stream volumetric flow rate and pollutant loading. In general, a small unit controlling a low pollutant loading will not be as cost effective as a large unit controlling a high pollutant loading. The costs presented are for flow rates of 470 m³/sec (1,000,000 scfm) and 1.0 m³/sec (2000 scfm), respectively, and a pollutant loading of 9 g/m³ (4.0 gr/ft³).

Pollutants that require an unusually high level of control or that require the fabric filter bags or the unit itself to be constructed of special materials, such as Gore-Tex or stainless steel, will increase the costs of the system. The additional costs for controlling more complex waste streams are not reflected in the estimates given below. For these types of systems, the capital cost could increase by as much as 75% and the operational and maintenance cost could increase by as much as 20%.

- Capital Cost: $13,100 to $54,900 per sm³/s ($6 to $26 per scfm)
- O&M Cost: $11,200 to $51,700 per sm³/s ($5 to $24 per scfm), annually
- Annualized Cost: $13,100 to $83,400 per sm³/s ($6 to $39 per scfm), annually
- Cost Effectiveness: $46 to $293 per metric ton ($42 to $266 per short ton)

In a fabric filter, flue gas is passed through a tightly woven or felted fabric, causing PM in the flue gas to be collected on the fabric by sieving and other mechanisms. Fabric filters may be in the form of sheets, cartridges, or bags, with a number of the individual fabric filter units housed together in a group. Bags are the most common type of fabric filter. The dust cake that forms on the filter from the collected PM can significantly increase collection efficiency. Fabric filters are frequently referred to as baghouses because the fabric is usually configured in cylindrical bags. Bags may be 6 to 9 m (20 to 30 ft) long and 12.7 to 30.5 cm (5 to 12 in) in diameter.

Groups of bags are placed in isolable compartments to allow cleaning of the bags or replacement of some of the bags without shutting down the entire fabric filter.

Operating conditions are important determinants of the choice of fabric. Some fabrics (e.g., polyolefins, nylons, acrylics, polyesters) are useful only at relatively low temperatures of 95 to 150°C (200 to 300°F). For high-temperature flue gas streams, more thermally stable fabrics such as fiberglass, Teflon, or Nomex must be used.

Practical application of fabric filters requires the use of a large fabric area in order to avoid an unacceptable pressure drop across the fabric. Baghouse size for a particular unit is determined by the choice of air-to-cloth ratio, or the ratio of volumetric airflow to cloth area. The selection of air-to-cloth ratio depends on the particulate loading and characteristics, and the cleaning method used. A high particulate loading will require the use of a larger baghouse in order to avoid forming too heavy a dust cake, which would result in an excessive pressure drop. As an example, a baghouse for a 250 MW utility boiler may have 5000 separate bags with a total fabric area approaching 46,500 m² (500,000 square feet).

Determinants of baghouse performance include the fabric chosen, the cleaning frequency and methods, and the particulate characteristics. Fabrics can be chosen that will intercept a greater fraction of particulate and some fabrics are coated with a membrane with very fine openings for enhanced removal of submicron particulate. Such fabrics tend to be more expensive.

Pulse jet cleaning of fabric filters is relatively new compared to other types of fabric filters. This cleaning mechanism has consistently grown in popularity because it can treat high dust loadings, operate at constant pressure drop, and occupy less space than other types of fabric filters. Pulse jet-cleaned fabric filters can only operate as external cake collection devices. The bags are closed at the bottom, open at the top, and supported by internal retainers, called cages. Particulate-laden gas flows into the bag, with diffusers often used to prevent oversized particles from damaging the bags. The gas flows from the outside to the inside of the bags, and then out the gas exhaust. The particles are collected on the outside of the bags and drop into a hopper below the fabric filter.

During pulse jet cleaning, a short burst, 0.03 to 0.1 seconds in duration, of high pressure (415 to 830 kPa) (60 to 120 pounds per square inch gauge (psig) of air is injected into the bags. The pulse is blown through a Venturi nozzle at the top of the bags and establishes a shockwave that continues onto the bottom of the bag. The wave flexes the fabric, pushing it away from the cage, and then snaps it back, dislodging the dust cake. The cleaning cycle is regulated by a remote timer connected to a solenoid valve. The burst of air is controlled by the solenoid valve and is released into blow pipes that have nozzles located above the bags. The bags are usually cleaned row by row.

There are several unique attributes of pulse jet cleaning. Because the cleaning pulse is very brief, the flow of dusty gas does not have to be stopped during cleaning. The other bags continue to filter, taking on extra duty because of the bags being cleaned. In general, there is no change in fabric filter pressure drop or performance as a result of pulse jet cleaning. This enables the pulse jet fabric filters to operate on a continuous basis, with solenoid valves as the only significant moving parts. Pulse jet cleaning is also more intense and occurs with greater frequency than the other fabric filter cleaning methods. This intense cleaning dislodges nearly all of the dust cake each time the bag is pulsed. As a result, pulse jet filters do not rely on a dust cake to provide filtration. Felted (non-woven) fabrics are used in pulse jet fabric filters because they do not require a dust cake to achieve high collection efficiencies. It has been found that woven fabrics used with pulse jet fabric filters leak a great deal of dust after they are cleaned.

Since bags cleaned by the pulse jet method do not need to be isolated for cleaning, pulse jet-cleaned fabric filters do not need extra compartments to maintain adequate filtration during cleaning. Also, because of the intense and frequent nature of the cleaning, they can treat higher gas flow rates with higher dust loadings. Consequently, fabric filters cleaned by the pulse jet method can be smaller than other types of fabric filters in the treatment of the same amount of gas and dust, making higher gas-to-cloth ratios achievable.

Fabric filters in general provide high collection efficiencies on both coarse and fine (submicron) particulate matter. They are relatively insensitive to fluctuations in gas stream conditions. Efficiency and pressure drop are relatively unaffected by large changes in inlet dust loadings for continuously cleaned filters. Filter outlet air is very clean and may be recirculated within the plant in many cases (for energy conservation). Material is collected dry for subsequent processing or disposal. Corrosion and rusting of components are usually not problems. Operation is relatively simple. Unlike electrostatic precipitators, fabric filter systems do not require the use of high voltage; therefore, maintenance is simplified, and flammable dust may be collected with proper care. The use of selected fibrous or granular filter aids (precoating) permits the high-efficiency collection of submicron smokes and gaseous contaminants. Filter collectors are available in a large number of configurations, resulting in a range of dimensions and inlet and outlet flange locations to suit installation requirements.

They do have some disadvantages. Temperatures much in excess of 290°C (550°F) require special refractory mineral or metallic fabrics, which can be expensive. Certain dusts may require fabric treatments to reduce dust seepage, or in other cases, assist in the removal of the collected dust. Concentrations of some dusts in the collector, approximately 50 g/m^3 (22 gr/ft^3), may represent a fire or explosion hazard if a spark or flame is accidentally admitted (i.e., risk of a dust explosion). Fabrics can burn if readily oxidizable dust is being collected. Fabric filters have relatively high maintenance requirements (e.g., periodic bag replacement). Fabric life maybe shortened at elevated temperatures and in the presence of acid or alkaline particulate or gas constituents. They cannot be operated in moist environments. Hygroscopic materials, condensation of moisture, or tarry adhesive components may cause crusty caking or plugging of the fabric or require special additives. Respiratory protection for maintenance personnel may be required when replacing fabric. Medium pressure drop is required, typically in the range of 100 to 250 mm of water column (4 to 10 inches of water column). A specific disadvantage of pulse jet units that use very high gas velocities is that the dust from the cleaned bags can be drawn immediately to the other bags. If this occurs, little of the dust falls into the hopper, and the dust layer on the bags becomes too thick. To prevent this, pulse jet fabric filters can be designed with separate compartments that can be isolated for cleaning.

Fabric filters are useful for collecting particles with resistivities either too low or too high for collection with electrostatic precipitators. Fabric filters therefore may be good candidates for collecting fly ash from low-sulfur coals or fly ash containing high unburned carbon levels, which respectively have high and low resistivities, and thus are relatively difficult to collect with electrostatic precipitators.

Dry Electrostatic Precipitator Wire-Pipe Type (ESP)

Typical new equipment design efficiencies are between 99 and 99.9%. Older existing equipment have a range of actual operating efficiencies of 90 to 99.9%. While several factors determine ESP collection efficiency, ESP size is most important. Size determines treatment time; the longer a particle spends in the ESP, the greater its chance of being collected. Maximizing electric field strength will maximize ESP collection efficiency. Collection efficiency is also affected by dust resistivity, gas temperature, chemical composition (of the dust and the gas), and particle size distribution. As with other equipment described, this hardware is applicable to point source emission control.

Many older ESPs are of the wire-pipe design, consisting of a single tube placed on top of a smokestack. Dry pipe-type ESPs are occasionally used by the textile industry, pulp and paper facilities, the metallurgical industry, coke ovens, hazardous waste incinerators, and sulfuric acid manufacturing plants, among others, though other ESP types are employed as well.

Wet wire-pipe ESPs are used much more frequently than dry wire-pipe ESPs, which are used only in cases in which wet cleaning is undesirable, such as high temperature streams or wastewater restrictions.

Figure 5.8 shows a commercial installation.

Typical gas flow rates for dry wire-pipe ESPs are 0.5 to 50 sm^3/sec (1000 to 100,000 scfm).

Dry wire-pipe ESPs can operate at very high temperatures, up to 700°C (1300°F). Operating gas temperature and chemical composition of the dust are key factors influencing dust resistivity and must be carefully considered in the design of an ESP.

Typical inlet concentrations to a wire-pipe ESP are 1 to 10 g/m^3 (0.5 to 5 gr/scfm). It is common to pretreat a waste stream, usually with a wet spray or scrubber, to bring the stream temperature and pollutant loading into a manageable range. Highly toxic flows with concentrations well below 1 g/m^3 (0.5 gr/scfm) are also sometimes controlled with ESPs.

In general, dry ESPs operate most efficiently with dust resistivities between 5 x 10^3 and 2 x 10^{10} ohm-cm. In general, the most difficult particles to collect are those with aerodynamic diameters between 0.1 and 1.0 µm. Particles between 0.2 and 0.4 µm usually show the most penetration. This is most likely a result of the transition region between field and diffusion charging.

When much of the loading consists of relatively large particles, mechanical collectors, such as cyclones or spray coolers, may be used to reduce the load on the ESP, especially at high inlet concentrations. Gas conditioning equipment to improve ESP performance by changing dust resistivity is occasionally used as part of the original design, but more frequently it is used to upgrade existing ESPs.

The equipment injects an agent into the gas stream ahead of the ESP. Usually, the agent mixes with the particles and alters their resistivity to promote higher migration velocity, and thus higher collection efficiency. Conditioning agents that are used include SO_3, H_2SO_4, sodium compounds, ammonia, and water; the conditioning agent most used is SO_3.

The following are cost ranges (expressed in third-quarter 1995 dollars) for dry wire-pipe ESPs of conventional design

Figure 5.8 Commerical installation.

under typical operating conditions, developed using EPA cost-estimating spreadsheets.[5] Costs can be substantially higher than in the ranges shown for pollutants, which require an unusually high level of control, or which require the ESP to be constructed of special materials such as stainless steel or titanium. In general, smaller units controlling a low concentration waste stream will not be as cost effective as a large unit cleaning a high pollutant load flow.

- Capital Cost: $65,000 to $400,000 per sm³/sec ($30 to $190 per scfm)
- O&M Cost: $10,000 to $20,000 per sm³/sec ($5 to $10 per scfm), annually
- Annualized Cost: $20,000 to $75,000 per sm³/sec ($10 to $35 per scfm), annually
- Cost Effectiveness: $55 to $950 per metric ton ($50 to $850 per short ton)

An ESP is a particulate control device that uses electrical forces to move particles entrained within an exhaust stream onto collection surfaces. The entrained particles are given an electrical charge when they pass through a corona, a region where gaseous ions flow. The power system that supplies the power for the electric field and the charging of the particles consists of two major components: (1) the transformer-rectifier set, and (2) the auto-voltage feedback control system. There may be one or more complete and independent power systems in a particular installation, depending on such factors as unit size, the characteristics of the effluent, and efficiency requirements.

Electrodes in the center of the flow lane are maintained at high voltage and generate the electrical field that forces the particles to the collector walls. In dry ESPs, the collectors are knocked, or "rapped," by various mechanical means to dislodge the particulate, which slides downward into a hopper where it is collected.

---

[5] EPA, 1996.

Recently, dry wire-pipe ESPs are being cleaned acoustically with sonic horns. The horns, typically cast metal horn bells, are usually powered by compressed air, and acoustic vibration is introduced by a vibrating metal plate that periodically interrupts the airflow. As with a rapping system, the collected particulate slides downward into the hopper. The hopper is evacuated periodically, as it becomes full. Dust is removed through a valve into a dust-handling system, such as a pneumatic conveyor, and is then disposed of in an appropriate manner.

In a wire-pipe ESP, also called a tubular ESP, the exhaust gas flows vertically through conductive tubes, generally with many tubes operating in parallel. The tubes may be formed as a circular, square, or hexagonal honeycomb. Square and hexagonal pipes can be packed closer together than cylindrical pipes, reducing wasted space. Pipes are generally 7 to 30 cm (3 to 12 in.) in diameter and 1 to 4 m (3 to 12 ft) in length. The high-voltage electrodes are long wires or rigid "masts" suspended from a frame in the upper part of the ESP that run through the axis of each tube. Rigid electrodes are generally supported by both an upper and lower frame. In modern designs, sharp points are added to the electrodes, either at the entrance to a tube or along the entire length in the form of stars, to provide additional ionization sites.

The power supplies for the ESP convert the industrial AC voltage (220 to 480 volts) to pulsating DC voltage in the range of 20,000 to 100,000 volts as needed. The voltage applied to the electrodes causes the gas between the electrodes to break down electrically, an action known as a "corona." The electrodes are usually given a negative polarity because a negative corona supports a higher voltage than does a positive corona before sparking occurs. The ions generated in the corona follow electric field lines from the electrode to the collection surfaces. Therefore, each electrode-pipe combination establishes a charging zone through which the particles must pass. As larger particles (>10 μm diameter) absorb many times more ions than small particles (>1 μm diameter), the electrical forces are much stronger on the large particles.

Due to necessary clearances needed for nonelectrified internal components at the top of wire-plate ESPs, part of the gas is able to flow around the charging zones. This is called "sneakage" and places an upper limit on the collection efficiency. Wire-pipe ESPs provide no sneakage paths around the collecting region, but field nonuniformities may allow some particles to avoid charging for a considerable fraction of the tube length. Dry wire-pipe ESPs are, however, subject to re-entrainment of the collected material after cleaning the collectors with a rapping or acoustic mechanism, though the closed nature of the pipes increases chances for recollection. The internal configuration of a conventional ESP is illustrated in Figure 5.9.

Another major factor in the performance is the resistivity of the collected material. Because the particles form a continuous layer on the ESP pipes, all the ion current must pass through the layer to reach the ground. This current creates an electric field in the layer, and it can become large enough to cause local electrical breakdown. When this occurs, new ions of the wrong polarity are injected into the wire-pipe gap, where they reduce the charge on the particles and may cause sparking. This breakdown condition is called "back corona." Back corona is prevalent when the resistivity of the layer is high, usually above 2×10^{11} ohm-cm. Above this level, the collection ability of the unit is reduced considerably because the severe back corona causes difficulties in charging the particles. Low resistivities will also cause problems. At resistivities below 10^8 ohm-cm, the particles are held on the collecting surface so loosely that general re-entrainment, as well as that associated with collector cleaning, becomes much more severe. Hence, care must be taken in measuring or estimating resistivity because it is strongly affected by such variables as temperature, moisture, gas composition, particle composition, and surface characteristics.

Dry wire-pipe ESPs and other ESPs in general, because they act only on the particulate to be removed, and only minimally hinder flue gas flow, have very low pressure drops (typically less than 13 mm (0.5 in.) water column). As a result, energy requirements and operating costs tend to be low. They

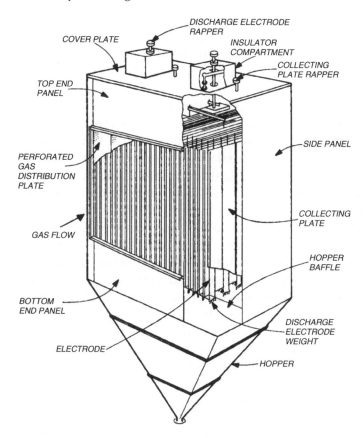

Figure 5.9 Internal configuration for a conventional ESP.

are capable of very high efficiencies, even for very small particles. They can be designed for a wide range of gas temperatures, and can handle high temperatures, up to 700°C (1300°F). Dry collection and disposal allows for easier handling. Operating costs are relatively low. ESPs are capable of operating under high pressure (to 1,030 kPa [150 psi]) or vacuum conditions. Relatively large gas flow rates can be effectively handled, though are uncommon in wire-pipe ESPs.

ESPs generally have high capital costs. Wire discharge electrodes (approximately 2.5 mm [0.01 in.] in diameter) are high-maintenance items. Corrosion can occur near the top of the wires because of air leakage and acid condensation. Also, long weighted wires tend to oscillate—the middle of the wire can approach the pipe, causing increased sparking and wear. Newer ESP designs are tending toward rigid electrodes, or "masts," which largely eliminate the drawbacks of using wire electrodes.

ESPs in general are not suited for use in processes that are highly variable because they are very sensitive to fluctuations in gas stream conditions (flow rates, temperatures, particulate and gas composition, and particulate loadings). ESPs are also difficult to install in sites that have limited space, since ESPs must be relatively large to obtain the low gas velocities necessary for efficient PM collection. Certain particulates are difficult to collect due to extremely high or low resistivity characteristics. There can be an explosion hazard when treating combustible gases or collecting combustible particulates. Relatively sophisticated maintenance personnel are required, as well as special precautions to safeguard personnel from the high voltage. Dry ESPs are not recommended for removing sticky or moist particles. Ozone is produced by the negatively charged electrode during gas ionization.[6]

Dusts with very high resistivities (greater than 10^{10} ohm-cm) are also not well-suited for collection in dry ESPs. These particles are not easily charged, and thus are not easily collected. High-resistivity particles also form ash layers with very high voltage gradients on the collecting electrodes. Electrical breakdowns in these ash layers lead to injection of positively charged ions into the space between the discharge and collecting electrodes (back corona), thus reducing the charge on particles in this space and lowering collection efficiency. Fly ash from the combustion of low-sulfur coal typically has a high resistivity, and thus is difficult to collect.

Wet Electrostatic Precipitator (ESP): Wire-Pipe Type and Others

Typical new equipment design efficiencies are between 99 and 99.9%. Older existing equipment have a range of actual

---

[6] AWMA, 1992.

operating efficiencies of 90 to 99.9%. Although several factors determine ESP collection efficiency, ESP size is most important. Size determines treatment time; the longer a particle spends in the ESP, the greater its chance of being collected. Maximizing electric field strength will maximize ESP collection efficiency. Collection efficiency is also affected to some extent by dust resistivity, gas temperature, chemical composition (of the dust and the gas), and particle size distribution.

Wet ESPs are used in situations for which dry ESPs are not suited, such as when the material to be collected is wet, sticky, flammable, or explosive, or has a high resistivity. Therefore, producer gas streams with tar represent a problem.

Also, as higher collection efficiencies have become more desirable, wet ESP applications have been increasing. Many older ESPs are of the wire-pipe design, consisting of a single tube placed on top of a smokestack. Wet pipe-type ESPs are commonly used by the textile industry, pulp and paper facilities, the metallurgical industry, coke ovens, hazardous waste incinerators, and sulfuric acid manufacturing plants, among others, though other ESP types are employed as well.

There are four types of precipitators that have used water in their operation:

- Intermittent-flush, parallel-plate precipitators
- Pipe-type precipitators
- Two-stage precipitators with intermittent flushing
- Continuous-spray, parallel-plate, wet electrostatic precipitators

The intermittent-flush, parallel-plate precipitator is built in either cylindrical or square housings and uses intermittent spraying to remove collected particles from the electrodes. The sprays are usually located just before or just after the precipitation fields. These sprays must not be applied at low pressure if the field strength is not to be interrupted. The intermittent-flush units are used on blast furnace applications.

The pipe-type units are usually built in a cylindrical housing having a header sheet near the inside top. In this header sheet are nested pipes, which act as collecting electrodes. The discharge electrodes are supported above the header sheet and hang axially in the collecting pipes. Water is introduced onto this sheet and flows over leveled weir rings, flushing the collection walls with a thin film of water.

Charged particles are collected in the water film, neutralized, and drained off at the bottom of the unit. This type of unit is commonly used on scarfing operations, detarring, and sulfuric acid mist collection. The two-stage precipitator is used where low ozone generation is required. It is frequently applied in the cleaning of recirculated ventilation. The particles are charged in the first field and collected in a second noncorona stage. Intermittent-flush mechanical flushing is used to clean these units. They are not well suited for external industrial applications because they are capable of accepting only very light loadings. Water is used only sometimes with this type of equipment, and therefore it has been classified as a type of wet precipitator.

The continuous-spray, parallel-plate design of precipitator is constructed with a rectangular housing. The sprays used in this system to clean the electrostatic collecting plates are located above the electrostatic field. The fine water droplets formed by the sprays are electrostatically deposited on the collecting surfaces. This system does not disrupt the electrostatic field power but does provide a better wetting action than does the intermittent-spray or weir-type unit. The wetting is continuous and uniform. The continuous flushing eliminates wet/dry build up problems experienced with other types of design. The weir-type systems do not distribute water evenly or continuously. Intermittent spray requires interruption of the gas cleaning cycle in most cases and can be used effectively only when deposit buildup is slow. The continuous wet precipitation has many broad and diverse applications. It is much more versatile and adaptable than its predecessors.

The continuous-spray design is ideal for any application in which there is a relatively light loading of submicron particles or condensed organics that form a submicron flame. Ordinarily, the only other piece of equipment applicable to this type of loading would be a high-energy scrubber. Since dust resistivity has no effect on the precipitator, it can be applied successfully on many very difficult dry applications.

Continuous-spray wet electrostatic precipitators have been applied on many applications, including:

- Soderberg aluminum reduction cells for simultaneous removal of aluminum oxides, solid and gaseous fluoride, tar mist (condensible hydrocarbons), and SO_2
- On carbon anode baking furnaces (ring furnaces) for removal of carbon particles, tar mists, and SO_2
- On fiberglass resin application sections and forming lines to remove short particles of glass fiber, phenolic resins, and tars
- On molybdenum sulfate roasting, downstream of a scrubber, to remove ammonium sulfite and sulfate aerosols that form in the ammonia scrubbing stage, and SO_2
- On phosphate rock driers for removal of submicron particulate and SO_2
- On coke batteries for the removal of fine carbon particles, condensible hydrocarbons, and SO_2

This precipitator is also applicable on all the following products:

- Acid mists and aerosols
- High-resistivity particulates
- Condensed particles or gaseous
- Oil- and tar-contaminated particles
- Oil mists and tar fogs

This particular design has a flexibility and range of applications that far exceed most other designs. Typical gas flow rates for wet wire-pipe ESPs are 0.5 to 50 sm^3/sec (1,000 to 100,000 scfm). Wet wire-pipe ESPs are limited to operating at temperatures lower than approximately 80 to 90°C (170 to 190°F).

Typical inlet concentrations to a wire-pipe ESP are 1 to 10 g/m^3 (0.5 to 5 gr/ft^3). It is common to pretreat a waste stream, usually with a wet spray or scrubber, to bring the stream temperature and pollutant loading into a manageable range. Highly toxic flows with concentrations well below 1 g/m^3 (0.5 gr/ft^3) are also sometimes controlled with ESPs.

When the gas loading is exceptionally high or consists of relatively large particles (> 2 µm), Venturi scrubbers or spray chambers may be used to reduce the load on the ESP. Much larger particles (> 10 µm) are controlled with mechanical collectors such as cyclones. Gas conditioning equipment to reduce both inlet concentration and gas temperature is occasionally used as part of the original design of wet ESPs.

The following are cost ranges (expressed in third-quarter 1995 dollars) for wire-pipe ESPs of conventional design under typical operating conditions, developed using EPA cost estimating spreadsheets for dry wire-pipe ESPs with adjustments made to reflect wet wire-pipe ESPs.[7] Costs can be substantially higher than in the ranges shown for pollutants, which require an unusually high level of control, or which require the ESP to be constructed of special materials such as titanium. Capital and operating costs are generally higher due to noncorrosive materials requirements, increased water usage, and treatment and disposal of wet effluent. In most cases, smaller units controlling a low concentration waste stream will not be as cost effective as a large unit cleaning a high pollutant load flow.[7]

- Capital Cost: $125,000 to $640,000 per sm^3/sec ($60 to $300 per scfm)
- O&M Cost: $15,000 to $25,000 per sm^3/sec ($7 to $12 per scfm), annually
- Annualized Cost: $32,000 to $125,000 per sm^3/s ($15 to $60 per scfm), annually
- Cost Effectiveness: $90 to $950 per metric ton ($80 to $850 per ton)

An ESP is a particulate control device that uses electrical forces to move particles entrained within an exhaust stream onto collection surfaces. The basic theory has already been described under dry ESPs, but a brief summary here is included, with notable differences between the equipment. The entrained particles are given an electrical charge when

---

[7] EPA, 1998.

they pass through a corona, a region where gaseous ions flow. Electrodes in the center of the flow lane are maintained at high voltage and generate the electrical field that forces the particles to the collector walls. In wet ESPs, the collectors are either intermittently or continuously washed by a spray of liquid, usually water. The collection hoppers used by dry ESPs are replaced with a drainage system. The wet effluent is collected, and often treated on-site.

In a wire-pipe ESP, also called a tubular ESP, the exhaust gas flows vertically through conductive tubes, generally with many tubes operating in parallel. The tubes may be formed as a circular, square, or hexagonal honeycomb. Square and hexagonal pipes can be packed closer together than cylindrical pipes, reducing wasted space. Pipes are generally 7 to 30 cm (3 to 12 in.) in diameter and 1 to 4 m (3 to 12 ft) in length. The high-voltage electrodes are long wires or rigid "masts" suspended from a frame in the upper part of the ESP that run through the axis of each tube. Rigid electrodes are generally supported by both an upper and lower frame. In modern designs, sharp points are added to the electrodes, either at the entrance to a tube or along the entire length in the form of stars, to provide additional ionization sites.

The power supplies for the ESP convert the industrial AC voltage (220 to 480 volts) to pulsating DC voltage in the range of 20,000 to 100,000 volts as needed. The voltage applied to the electrodes causes the gas between the electrodes to break down electrically, an action known as a "corona." The electrodes are usually given a negative polarity because a negative corona supports a higher voltage than does a positive corona before sparking occurs. The ions generated in the corona follow electric field lines from the electrode to the collecting pipe. Therefore, each electrode-pipe combination establishes a charging zone through which the particles must pass. As larger particles (> 10 μm diameter) absorb many times more ions than small particles (> 1 μm diameter), the electrical forces are much stronger on the large particles.

Due to necessary clearances needed for nonelectrified internal components at the top of wire-pipe ESPs, part of the gas is able to flow around the charging zones. This is called "sneakage" and places an upper limit on the collection efficiency. Wire-pipe ESPs provide no sneakage paths around the collecting region, but field nonuniformities may allow some particles to avoid charging for a considerable fraction of the tube length. Wet ESPs require a source of wash water to be injected or sprayed near the top of the collector pipes either continuously or at timed intervals. This wash system replaces the rapping mechanism usually used by dry ESPs. The water flows with the collected particles into a sump from which the fluid is pumped or drained. A portion of the fluid may be recycled to reduce the total amount of water required. The remainder is pumped into a settling pond or passed through a dewatering stage, with subsequent disposal of the sludge.

Unlike dry ESPs, resistivity of the collected material is generally not a major factor in performance. Because of the high humidity in a wet ESP, the resistivity of particles is lowered, eliminating the "back corona" condition. The frequent washing of the pipes also limits particle buildup on the collectors.

The power consumed to operate a wet electrostatic precipitator is much less than that required by most other methods of control. There are four areas in which power is consumed: (1) electrostatic power, (2) fan power, (3) insulator heating power, and (4) pump power. The total electrostatic power input required for operation is 0.8 to 1.0 kW/1000 ft^3 of collection area. A comparable piece of equipment is a Venturi scrubber with 50-in.wg pressure drop. The power required for this installation would be 6 to 7 kW/1000 cfm. This would mean that approximately seven times the power would be needed to achieve the same amount of cleaning with a Venturi scrubber as opposed to using a precipitator.

Since power is a substantial component of the fixed operating cost of a unit, the operating cost would run approximately seven times more on a scrubber installation. The installation costs of a hot-rolled steel precipitator to handle 100,000 cfm would be between $3.50 and $4.50/cfm as opposed to $1.40 and $1.80/cfm for a Venturi scrubbing system.

Although the initial capital expenditure is high for the precipitator, if the total operating and capital costs are amortized over 8 to 10 years, the precipitator will prove to be the more economically feasible choice because of its low operating and maintenance costs.

The advantages of this equipment are as follows. Wet wire-pipe ESPs and other ESPs in general, because they act only on the particulate to be removed, and only minimally hinder flue gas flow, have very low pressure drops (typically less than 13 mm [0.5 in.] water column). As a result, energy requirements and operating costs tend to be low. They are capable of very high efficiencies, even for very small particles. Operating costs are relatively low. ESPs are capable of operating under high pressure (to 1030 kPa [150 psi]) or vacuum conditions, and relatively large gas flow rates can be effectively handled.

Another advantage is that wet ESPs can collect sticky particles and mists, as well as highly resistive or explosive dusts. The continuous or intermittent washing with a liquid eliminates the re-entrainment of particles due to rapping, which dry ESPs are subject to. The humid atmosphere that results from the washing in a wet ESP enables them to collect high resistivity particles, absorb gases or cause pollutants to condense, and cools and conditions the gas stream. Liquid particles or aerosols present in the gas stream are collected along with particles and provide another means of rinsing the collection electrodes.

Wet wire-pipe ESPs have the additional advantages of reducing "sneakage" by passing the entire gas stream through the collection field, and the ability to be tightly sealed to prevent leaks of material, especially valuable or hazardous materials.

Disadvantages are as follows: ESPs generally have high capital costs. Wire discharge electrodes (approximately 2.5 mm 0.01 in] in diameter) are high-maintenance items. Corrosion can occur near the top of the wires because of air leakage and acid condensation. Also, long weighted wires tend to oscillate — the middle of the wire can approach the pipe, causing increased sparking and wear. Newer ESP designs are tending toward rigid electrodes, or "masts," which largely eliminate the drawbacks of using wire electrodes (see Cooper and Alley, 1994; Flynn, 1999).

ESPs in general are not suited for use in processes that are highly variable because they are very sensitive to fluctuations in gas stream conditions (flow rates, temperatures, particulate and gas composition, and particulate loadings). ESPs are also difficult to install in sites that have limited space since ESPs must be relatively large to obtain the low gas velocities necessary for efficient PM collection. Relatively sophisticated maintenance personnel are required, as well as special precautions to safeguard personnel from the high voltage. Ozone is produced by the negatively charged electrode during gas ionization.

Wet ESPs add to the complexity of a wash system, because of the fact that the resulting slurry must be handled more carefully than a dry product, and in many cases requires treatment, especially if the dust can be sold or recycled. Wet ESPs are limited to operating at stream temperatures under approximately 80 to 90°C (170 to 190°F), and generally must be constructed of noncorrosive materials.

For wet ESPs, consideration must be given to handling wastewaters. For simple systems with innocuous dusts, water with particles collected by the ESP may be discharged from the ESP system to a solids-removing clarifier (either dedicated to the ESP or part of the plant wastewater treatment system) and then to final disposal. More complicated systems may require skimming and sludge removal, clarification in dedicated equipment, pH adjustment, or treatment to remove dissolved solids. Spray water from an ESP preconditioner may be treated separately from the water used to wash the ESP collecting pipes so that the cleaner of the two treated water streams may be returned to the ESP. Recirculation of treated water to the ESP may approach 100% (AWMA, 1992).

Venturi Scrubbers

This type of technology is a part of the group of air pollution controls collectively referred to as "wet scrubbers." Venturi scrubbers are also known as Venturi jet scrubbers, gas-atom-izing spray scrubbers, and ejector-Venturi scrubbers. The technology is based on the removal of air pollutants by inertial and diffusional interception.

Venturi scrubbers are primarily used to control particu-late matter, including PM less than or equal to 10 μm in aerodynamic diameter (PM$_{10}$), and PM less than or equal to 2.5 μm in aerodynamic diameter (PM$_{2.5}$). Though capable of some incidental control of volatile organic compounds (VOCs), generally Venturi scrubbers are limited to control PM and high solubility gases.

Venturi scrubbers PM collection efficiencies range from 70 to greater than 99%, depending upon the application. Collec-tion efficiencies are generally higher for PM with aerodynamic diameters of approximately 0.5 to 5 gm. Some Venturi scrub-bers are designed with an adjustable throat to control the velocity of the gas stream and the pressure drop. Increasing the Venturi scrubber efficiency requires increasing the pressure drop, which, in turn, increases the energy consumption.

Venturi scrubbers have been applied to control PM emis-sions from utility, industrial, commercial, and institutional boilers fired with coal, oil, wood, and liquid waste. They have also been applied to control emission sources in the chemical, mineral products, wood, pulp and paper, rock products, and asphalt manufacturing industries; lead, aluminum, iron and steel, and gray iron production industries; and municipal solid waste incinerators. Typically, Venturi scrubbers are applied where it is necessary to obtain high collection efficiencies for fine PM. Thus, they are applicable to controlling emission sources with high concentrations of submicron PM.

Typical gas flow rates for a single-throat Venturi scrub-ber unit are 0.2 to 28 sm³/sec (500 to 60,000 scfm). Flows higher than this range use either multiple Venturi scrubbers in parallel or a multiple throated Venturi.

Inlet gas temperatures are usually in the range of 4 to 370°C (40 to 700°F). Gas pollutant loadings can range from 1 to 115 g/sm³ (0.1 to 50 gr/scf).

Generally, no pretreatment is required for Venturi scrub-bers, though in some cases the gas is quenched to reduce the temperature for scrubbers made of materials affected by high temperatures.

The following are cost ranges (expressed in third-quarter 1995 dollars) for Venturi wet scrubbers of conventional design under typical operating conditions, developed using EPA cost estimating spreadsheets (EPA, 1996) and referenced to the volumetric flow rate of the waste stream treated. For purposes of calculating the example cost effectiveness, the pollutant is assumed to be PM at an inlet loading of approximately 7 g/sm^3 (3 gr/scf). The costs do not include costs for post-treatment or disposal of used solvent or waste. Actual costs can be substantially higher than in the ranges shown for applications that require expensive materials, solvents, or treatment methods. As a rule, smaller units controlling a low concentration waste stream will be much more expensive (per unit volumetric flow rate) than a large unit cleaning a high pollutant load flow.

- Capital Cost: $6,700 to $59,000 per sm^3/sec ($3.20 to $28 per scfm)
- O&M Cost: $8,700 to $250,000 per sm^3/sec ($4.10 to $119 per scfm), annually
- Annualized Cost: $9,700 to $260,000 per sm^3/sec ($4.60 to $123 per scfm), annually
- Cost Effectiveness: $84 to $2,300 per metric ton ($76 to $2,100 per short ton), annualized cost per ton per year of pollutant controlled

A Venturi scrubber accelerates the waste gas stream to atomize the scrubbing liquid and to improve gas-liquid contact. In a Venturi scrubber, a "throat" section is built into the duct that forces the gas stream to accelerate as the duct narrows and then expands (see Figure 5.10). As the gas enters the Venturi throat, both gas velocity and turbulence increase.

Depending on the scrubber design, the scrubbing liquid is sprayed into the gas stream before the gas encounters the Venturi throat, or in the throat, or upwards against the gas flow in the throat. The scrubbing liquid is then atomized into small droplets by the turbulence in the throat, and droplet-particle interaction is increased. Some designs use supplemental hydraulically or pneumatically atomized sprays to augment

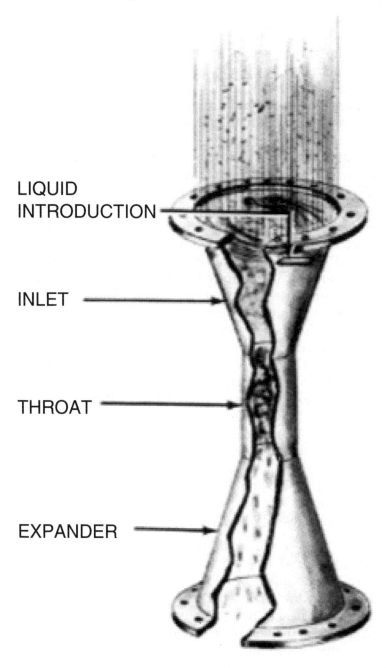

LIQUID
INTRODUCTION

INLET

THROAT

EXPANDER

Figure 5.10 Venturi scrubber details.

droplet creation. The disadvantage of these designs is that clean liquid feed is required to avoid clogging.

After the throat section, the mixture decelerates, and further impacts occur causing the droplets to agglomerate. Once the particles have been captured by the liquid, the wetted PM and excess liquid droplets are separated from the gas stream by an entrainment section, which usually consists of a cyclonic separator or a mist eliminator.

Current designs for Venturi scrubbers generally use the vertical downflow of gas through the Venturi throat and incorporate three features: (1) a "wet-approach" or "flooded-wall" entry section to avoid a dust buildup at a wet-dry junction; (2) an adjustable throat for the Venturi throat to provide for adjustment of the gas velocity and the pressure drop; and (3) a "flooded" elbow located below the Venturi and ahead of the entrainment separator, to reduce wear by abrasive particles. The Venturi throat is sometimes fitted with a refractory lining to resist abrasion by dust particles. Some examples of other configurations are shown in Figures 5.11 through 5.14. Figure 5.11 illustrates a rectangular Venturi scrubber configuration. Figure 5.12 shows a variable cylindrical Venturi scrubber design. Still another example is a nonwetted approach, variable rectangular Venturi scrubber shown in Figure 5.13. A wetted approach, variable rectangular Venturi scrubber is shown in Figure 5.14. Specific configurations are oftentimes established based upon available floor and headspace. As already noted, this is a versatile control technology with the dual function of removing PM and highly soluble gaseous pollutants. Some VOCs fall into the soluble gas category.

The advantages of Venturi scrubbers include:

- Capable of handling flammable and explosive dusts with little risk
- Can handle mists
- Relatively low maintenance
- Simple in design and easy to install
- Collection efficiency can be varied
- Provides cooling for hot gases
- Corrosive gases and dusts can be neutralized

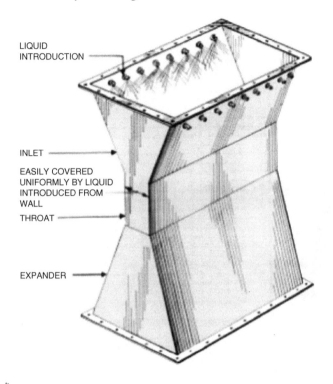

LIQUID
INTRODUCTION

INLET

EASILY COVERED
UNIFORMLY BY LIQUID
INTRODUCED FROM
WALL

THROAT

EXPANDER

Figure 5.11 Rectangular Venturi scrubbers.

Disadvantages of impingement plate scrubbers include:

- Effluent liquid can create water pollution problems
- Waste product collected wet
- High potential for corrosion problems
- Protection against freezing required
- Off-gas may require reheating to avoid visible plume
- Collected PM may be contaminated and may not be recyclable
- Disposal of waste sludge may be very expensive

For PM applications, wet scrubbers generate waste in the form of a slurry or wet sludge. This creates the need for both wastewater treatment and solid waste disposal. Initially, the slurry is treated to separate the solid waste from the

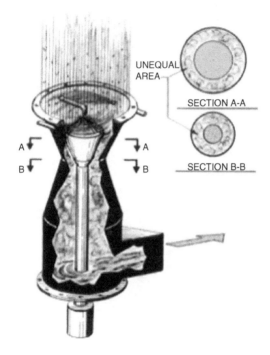

UNEQUAL
AREA

SECTION A-A

A

A

B

B

SECTION B-B

Figure 5.12   Variable cyclindrical Venturi scrubber configuration.

water. The treated water can then be reused or discharged. Once the water is removed, the remaining waste will be in the form of a solid or sludge.

If the solid waste is inert and nontoxic, it can generally be landfilled. Hazardous wastes will have more stringent procedures for disposal. In some cases, the solid waste may have value and can be sold or recycled.

Orifice Scrubber

This type of technology is a part of the group of controls collectively referred to as "wet scrubbers." Orifice scrubbers are also known as self-induced spray scrubbers, gas-induced spray scrubbers, and entrainment scrubbers. The operating

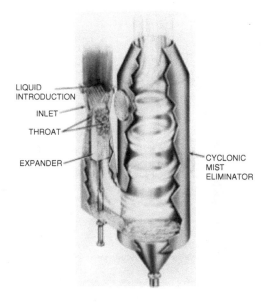

LIQUID INTRODUCTION

INLET

THROAT

EXPANDER

CYCLONIC MIST ELIMINATOR

Figure 5.13 Non-wetted design configuration.

principle is based on the removal of air pollutants by inertial and diffusional interception.

Orifice scrubbers are primarily used to control particulate matter, including particulate matter less than or equal to 10 μm in aerodynamic diameter, particulate matter less than or equal to 2.5 μm in aerodynamic diameter, down to particles with an aerodynamic diameter of approximately 2 μm.

Orifice scrubber collection efficiencies range from 80 to 99%, depending upon the application and scrubber design. This type of scrubber relies on inertial and diffusional interception for PM collection. Some orifice scrubbers are designed with adjustable orifices to control the velocity of the gas stream.

Orifice scrubbers are used in industrial applications including food processing and packaging (cereal, flour, rice, salt, sugar); pharmaceutical processing and packaging; and the manufacture of chemicals, rubber and plastics, ceramics, and fertilizer. Processes controlled include dryers, cookers,

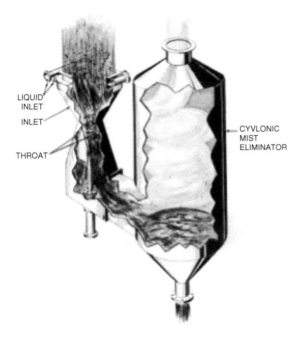

LIQUID
INLET

INLET

THROAT

CYVLONIC
MIST
ELIMINATOR

Figure 5.14 Wetted approach, rectangular design.

crushing and grinding operations, spraying (pill coating, ceramic glazing), ventilation (bin vents, dumping operations), and material handling (transfer stations, mixing, dumping, packaging). Orifice scrubbers can be built as high-energy units, but most devices are designed for low-energy service.

Typical gas flow rates for an orifice scrubber unit are 0.47 to 24 sm³/sec (1,000 to 50,000 scfm). In general, orifice scrubbers can treat waste gases up to approximately 150°C (300°F).

Orifice scrubbers can accept gas flows with PM loadings up to 23 g/sm³ or 10 gr/scf, or higher, depending upon the nature of the PM being controlled.

Orifice scrubbers generally do not require precleaning, unless the waste gas contains large pieces of debris. Precooling may be necessary for high-temperature waste gas flows, which increase the evaporation of the scrubbing liquid.

The following are cost ranges (expressed in third-quarter 1995 dollars) for orifice wet scrubbers of conventional design under typical operating conditions, adapted from EPA cost estimating spreadsheets (EPA, 1996) and referenced to the volumetric flow rate of the waste stream treated. For purposes of calculating the example cost effectiveness, the pollutant is PM at a loading of approximately 7 g/sm^3 (3 gr/scf). The costs do not include costs for post-treatment or disposal of used solvent or waste. Costs can be higher than in the ranges shown for applications that require expensive materials, solvents, or treatment methods. As a rule, smaller units controlling a low concentration waste stream will be more expensive (per unit volumetric flow rate) than a large unit cleaning a high pollutant load flow.

- Capital Cost: $10,000 to $36,000 per sm^3/sec ($5.00 to $17 per scfm)
- O&M Cost: $8,000 to $149,000 per sm^3/sec ($3.80 to $70 per scfm), annually
- Annualized Cost: $9,500 to $154,000 per sm^3/sec ($4.50 to $73 per scfm), annually
- Cost Effectiveness: $88 to $1,400 per metric ton ($80 to $1,300 per short ton), annualized cost per ton per year of pollutant controlled

Orifice scrubbers form a category of gas-atomized spray scrubbers in which a tube or a duct of some other shape forms the gas-liquid contacting zone. The particle-laden gas stream is forced to pass over the surface of a pool of scrubbing liquid at high velocity, entraining it as droplets as it enters an orifice. The gas stream flowing through the orifice atomizes the entrained liquid droplets in essentially the same manner as a Venturi scrubber. As the gas velocity and turbulence increases with the passing of the gas through the narrow orifice, the interaction between the PM and atomized liquid droplets also increases. Particulate matter and droplets are then removed from the gas stream by impingement on a series of baffles that the gas stream encounters after exiting the orifice. The collected liquid and PM drain from the baffles back into the liquid pool below the orifice.

The scrubbing liquid is fed into the pool at the bottom of the scrubber and later recirculated from the entrainment separator baffles by gravity instead of being circulated by a pump, as in Venturi scrubbers. Many devices using contactor ducts of various shapes are offered commercially. The principal advantage of this scrubber is the elimination of a pump for recirculation of the scrubbing liquid.

Advantages of orifice scrubbers include:

- Can handle flammable and explosive dusts with little risk
- Can handle mists
- Relatively low water recirculation rate
- Collection efficiency can be varied
- Provides cooling for hot gases
- Corrosive gases and dusts can be neutralized

Disadvantages of orifice scrubbers include:

- Effluent liquid can create water pollution problems
- Waste product collected wet
- High potential for corrosion problems
- Protection against freezing required
- Off-gas may require reheating to avoid visible plume
- Collected PM may be contaminated and may not be recyclable
- Disposal of waste sludge may be very expensive

For PM applications, wet scrubbers generate waste in the form of a slurry or wet sludge. This creates the need for both wastewater treatment and solid waste disposal. Initially, the slurry is treated to separate the solid waste from the water. The treated water can then be reused or discharged. Once the water is removed, the remaining waste will be in the form of a solid or sludge. If the solid waste is inert and nontoxic, it can generally be landfilled. Hazardous wastes will have more stringent procedures for disposal. In some cases, the solid waste may have value and can be sold or recycled.

Orifice scrubbers usually have low liquid demands, since they use the same scrubbing liquid for extended periods of time. Because orifice scrubbers are relatively simple in design and usually have few moving parts, the major maintenance concern is the removal of the sludge that collects at the bottom of the scrubber. Orifice scrubbers rarely drain continually from the bottom, because a static pool of scrubbing liquid is needed at all times. Therefore, the sludge is usually removed with a sludge ejector that operates like a conveyer belt. As the sludge settles to the bottom of the scrubber, it lands on the ejector and is conveyed up and out of the scrubber.

Orifice scrubbers are relatively simple in design and usually have few moving parts, aside from a fan and possibly an automatic sludge ejector.

Condensation Scrubbers

This type of technology is a part of the group of air pollution controls collectively referred to as "wet scrubbers." The removal of air pollutants is achieved by the use of condensation to increase pollutant particle size, followed by inertial interception. Condensation scrubbers are typically intended to control fine PM with an aerodynamic diameter of between approximately 0.25 and 1.0 µm.

Collection efficiencies of greater than 99% have been reported for particulate emissions, based on study results. Note list of references cited at the end of this chapter for detailed studies.

Condensation scrubbers are intended for use in controlling fine PM-containing waste-gas streams, and are designed specifically to capture fine PM that has escaped a primary PM control device. Condensation scrubbing systems are a relatively new technology and are not yet generally commercially available. It may be argued that this is a pollution prevention type of technology since it replaces other approaches to controlling very fine PM, although the primary role is end-of-pipe treatment.

Typical gas flows are on the order of 10 sm^3/sec or 21,000 scfm.

The gas entering a condensation scrubber is generally cooled to saturation conditions, approximately 20 to 26°C (68 to 78°F).

Gas loading is dependent upon the control effectiveness for fine PM of the primary PM control system. Fine PM may, in some cases, comprise up to 90% of the total mass of PM emissions from a combustion source, and many primary control technologies have relatively low collection efficiencies for fine PM.

For PM control from combustion sources, the flue gas enters a coagulation area (e.g., ductwork, a chamber, or a cyclone) to reduce the number of ultrafine particles, and then a gas conditioner to cool the gas to a suitable temperature and saturation state. This is generally accomplished by means of a waste heat recovery heat exchanger to reduce the temperature of the flue gas or by spraying water directly into the hot flue gas stream.

It is usually not practical or cost effective to cool flue gases to temperatures below ambient values. Condensation scrubbers are generally intended to be used downstream of another scrubber (e.g., a Venturi scrubber) that has already removed PM > 1.0 μm aerodynamic diameter.

The following provides cost information (expressed in fourth-quarter 1993 dollars) for retrofitting an existing scrubber system with a condensation scrubber under typical operating conditions, adapted from EPA cost-estimating spreadsheets (EPA, 1996) and referenced to the volumetric flow rate of the waste stream treated. For purposes of calculating the example cost effectiveness, the pollutant is PM at a loading of approximately 7 g/sm^3 or 3 gr/scf. The costs do not include costs for post-treatment or disposal of used solvent or waste.

- Capital Cost: $13,000 per sm^3/sec ($6.00 per scfm)
- O&M Cost: $5,300 per sm^3/sec ($2.50 per scfm), annually
- Annualized Cost: $7,000 per sm^3/sec ($3.40 per scfm), annually
- Cost Effectiveness: $65 per metric ton ($59 per short ton), annualized cost per ton per year of pollutant controlled

Condensation scrubbing is a relatively recent development in wet scrubber technology. Most conventional scrubbers rely on the mechanisms of impaction and diffusion to achieve contact between the PM and liquid droplets. In a condensation scrubber, the PM acts as condensation nuclei for the formation of droplets. Generally, condensation scrubbing depends on first establishing saturation conditions in the gas stream. Once saturation is achieved, steam is injected into the gas stream. The steam creates a condition of supersaturation and leads to condensation of water on the fine PM in the gas stream. The large condensed droplets are then removed by one of several conventional devices, such as a high-efficiency mist eliminator.

Advantages of condensation scrubbers include:

- Can handle flammable and explosive dusts with little risk
- Can handle fine PM
- Collection efficiency can be varied
- Corrosive gases and dusts can be neutralized

Disadvantages of condensation scrubbers include:

- Effluent liquid can create water pollution problems
- Waste product collected wet
- High potential for corrosion problems
- Protection against freezing required
- Off-gas may require reheating to avoid visible plume
- Collected particulate may be contaminated, and may not be recyclable
- Disposal of waste sludge may be very expensive

For PM applications, wet scrubbers generate waste in the form of a slurry. This creates the need for both wastewater treatment and solid waste disposal. Initially, the slurry is treated to separate the solid waste from the water. The treated water can then be reused or discharged.

GAS CONDITIONING TECHNOLOGIES

An overview of several technologies is presented in this section. Each of these technologies are suitable for handling gaseous pollutants. They are all well-established methods of gas cleaning, some based on many decades of industry practice.

Packed Tower and Absorption

This type of technology is a part of the group of controls collectively referred to as wet scrubbers. When used to control inorganic gases, they may also be referred to as "acid gas scrubbers." The technology is based on the removal of air pollutants by inertial or diffusional impaction, reaction with a sorbent or reagent slurry, or absorption into liquid solvent. This is an older technology, with much of the classical development work having been done in the 1920s and 30s. It is a very important technology, having been applied so widely, with both standard and newer applications still evolving.

Scrubbers used in air pollution control are available in a wide range of types and sizes. They are used mainly to eliminate one or more objectionable gaseous or particulate components from a gas stream. Their principle of design is based on mass transfer (diffusion), inertial impaction, or electrostatic attraction. In mass transfer, gaseous components are dissolved in liquid. When gas and liquid are brought into intimate contact, the concentration gradient is established between two phases, and diffusion takes place. Objectionable components, higher in concentration in the gaseous phase, are transferred to the liquid phase, having a lower concentration. During this diffusion process, solute is transferred in liquid with or without chemical reaction.

When fluid passes over a solid surface, its velocity at the surface of the solid is 0. The velocity of the fluid thus changes from that in the bulk stream to the solid surface across which it is flowing. The velocity rises sharply in a zone between the interface and a very small distance perpendicular to the solid interface. This small zone is called the laminar region. The fluid in the bulk stream can be in the turbulent region. The zone between the laminar and turbulent regions is known as the transition or buffer region. Intimate contact between gas and liquid is established in the laminar region over solids known as packing. The laminar region consists of stagnant gas and liquid films. As diffusing fluid passes from the main stream, it has to pass through the main stream, buffer zone, and laminar regions.

Diffusion through the laminar film is on a molecular scale and is known as molecular diffusion. Molecules in gases move in random directions, and as they do so, they collide against each other. The resultant distance is, therefore, very small; hence, molecular diffusion is a slow process. If, on the other hand, the temperature of the gas is higher, molecules travel at a higher velocity and can cover larger distances, thereby increasing the rate of diffusion. At lower pressure, there is a greater distance between molecules; this can also increase the rate of diffusion.

The main concentration gradient is established in the laminar region, so the mechanism of gas absorption by diffusion is a molecular diffusion.

The technology is primarily applicable to the removal of inorganic fumes, vapors, and gases (e.g., chromic acid, hydrogen sulfide, ammonia, chlorides, fluorides, and SO_2); volatile organic compounds; and particulate matter, including PM less than or equal to 10 μm in aerodynamic diameter (PM_{10}), PM less than or equal to 2.5 μm in aerodynamic diameter ($PM_{2.5}$), and hazardous air pollutants (HAP) in particulate form (PM_{HAP}).

Absorption is widely used as a raw material or product recovery technique in separation and purification of gaseous streams containing high concentrations of VOC, especially water-soluble compounds such as methanol, ethanol, isopropanol, butanol, acetone, and formaldehyde. Hydrophobic VOC can be absorbed using an amphiphilic block copolymer dissolved in water. However, as an emission control technique, it is much more commonly employed for controlling inorganic gases than for VOC. When using absorption as the primary control technique for organic vapors, the spent solvent must be easily regenerated or disposed of in an environmentally acceptable manner. When used for PM control, high concentrations can clog the bed, limiting these devices to controlling streams with relatively low dust loadings.

Control device vendors estimate that removal efficiencies range from 95 to 99%.

Removal efficiencies for gas absorbers vary for each pollutant-solvent system and with the type of absorber used. Most absorbers have removal efficiencies in excess of 90%, and packed-tower absorbers may achieve efficiencies greater than 99% for some pollutant-solvent systems. The typical collection efficiency range is from 70 to greater than 99%.

Packed-bed wet scrubbers are limited to applications in which dust loading is low, and collection efficiencies range from 50 to 95%, depending upon the application. Condensation scrubbers potentially offer a means of extending the removal efficiency of PM.

The suitability of gas absorption as a pollution control method is generally dependent on the following factors: (1) availability of suitable solvent; (2) required removal efficiency; (3) pollutant concentration in the inlet vapor; (4) capacity required for handling waste gas; and (5) recovery value of the pollutant(s) or the disposal cost of the unrecoverable solvent. Packed-bed scrubbers are typically used in the chemical, aluminum, coke and ferro-alloy, food and agriculture, and chromium electroplating industries.

These scrubbers have had limited use as part of flue gas desulfurization (FGD) systems, but the scrubbing solution flow rate must be carefully controlled to avoid flooding. When absorption is used for VOC control, packed towers are usually more cost effective than impingement plate towers.

However, in certain cases, the impingement plate design is preferred over packed-tower columns either when internal cooling is desired or where low liquid flow rates would inadequately wet the packing.

Typical gas flow rates for packed-bed wet scrubbers are 0.25 to 35 sm^3/sec (500 to 75,000 scfm).

Inlet temperatures are usually in the range of 4 to 370°C (40 to 700°F) for waste gases in which the PM is to be controlled, and for gas absorption applications, 4 to 38°C (40 to 100°F). In general, the higher the gas temperature, the lower the absorption rate, and vice versa. Excessively high gas temperatures also can lead to significant solvent or scrubbing liquid loss through evaporation.

Typical gaseous pollutant concentrations range from 250 to 10,000 parts per millions by volume (ppmv). Packed-bed wet scrubbers are generally limited to applications in which PM concentrations are less than 0.45 g/sm^3 (0.20 gr/scf) to avoid clogging.

For absorption applications, precoolers (e.g., spray chambers, quenchers) may be needed to saturate the gas stream or to reduce the inlet air temperature to acceptable levels to avoid solvent evaporation or reduced absorption rates.

The following are cost ranges (expressed in third-quarter 1995 dollars) for packed-bed wet scrubbers of conventional design under typical operating conditions, developed using EPA cost estimating spreadsheets (EPA, 1996a) and referenced to the volumetric flow rate of the waste stream treated. For purposes of calculating the example cost effectiveness, the pollutant used is hydrochloric acid and the solvent is aqueous caustic soda. The costs do not include costs for post-treatment or disposal of used solvent or waste. Costs can be substantially higher than in the ranges shown for applications that require expensive materials, solvents, or treatment methods. As a rule, smaller units controlling a low concentration waste stream will be much more expensive (per unit volumetric flow rate) than a large unit cleaning a high pollutant load flow.

- Capital Cost: $22,500 to $120,000 per sm^3/sec ($11 to $56 per scfm)
- O&M Cost: $33,500 to $153,000 per sm^3/sec ($16 to $72 per scfm), annually
- Annualized Cost: $36,000 to $166,000 per sm^3/sec ($17 to $78 per scfm), annually
- Cost Effectiveness: $0.24 to $1.09 per metric ton ($0.21 to $0.99 per short ton), annualized cost per ton per year of pollutant controlled

Figures 5.15 through 5.18 show examples of different scrubber configurations and operating modes.

Physical absorption depends on properties of the gas stream and liquid solvent, such as density and viscosity, as well as specific characteristics of the contaminants in the gas and the liquid stream (e.g., diffusivity, equilibrium solubility).

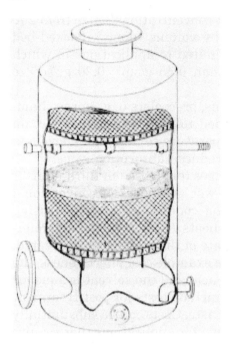

Figure 5.15 Countercurrent packed scrubber.

These properties are temperature dependent, and lower temperatures generally favor absorption of gases by the solvent. Absorption is also enhanced by greater contacting surface, higher liquid-gas ratios, and higher concentrations in the gas stream. Chemical absorption may be limited by the rate of reaction, although the rate-limiting step is typically the physical absorption rate, not the chemical reaction rate.

Water is the most common solvent used to remove inorganic contaminants. Contaminant removal may be enhanced by manipulating the chemistry of the absorbing solution so that it reacts with the pollutant. Caustic solution (sodium hydroxide, NaOH) is the most common scrubbing liquid used for acid-gas control (e.g., HCl, SO_2, or both), though sodium carbonate (Na_2CO_3) and calcium hydroxide (slaked lime, $Ca[OH]_2$) are also used. When the acid gases are absorbed into the scrubbing solution, they react with alkaline compounds to produce neutral salts. The rate of absorption of the

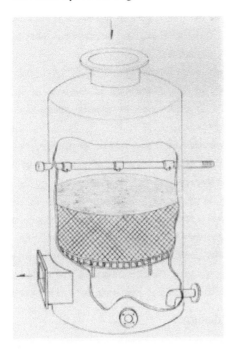

Figure 5.16 Cocurrent packed scrubber design.

acid gases is dependent upon the solubility of the acid gases in the scrubbing liquid.

Absorption is a commonly applied operation in chemical processing. It is used as a raw material or a product recovery technique in separation and purification of gaseous streams containing high concentrations of organics (e.g., in natural gas purification and coke by-product recovery operations). In absorption, the organics in the gas stream are dissolved in a liquid solvent. The contact between the absorbing liquid and the vent gas is accomplished in countercurrent spray towers, scrubbers, or packed or plate columns.

The use of absorption as the primary control technique for organic vapors is subject to several limiting factors. One factor is the availability of a suitable solvent. The VOC must be soluble in the absorbing liquid and even then, for any given absorbent liquid, only VOCs that are soluble can be removed.

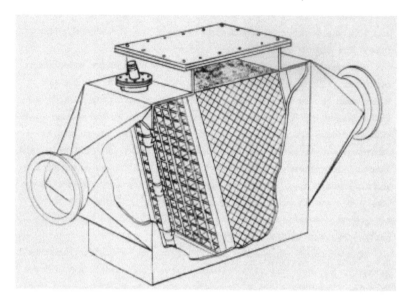

Figure 5.17   Parallel scrubber design.

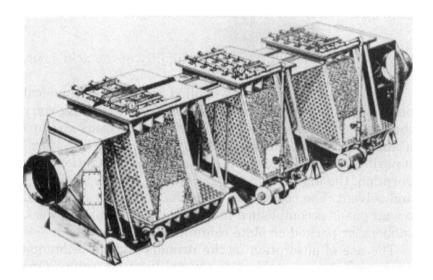

Figure 5.18   Cross-flow scrubber design.

Some common solvents that may be useful for volatile organics include water, mineral oils, or other nonvolatile petroleum oils. Another factor that affects the suitability of absorption for organic emissions control is the availability of vapor/liquid equilibrium data for the specific organic/solvent system in question. Such data are necessary for the design of absorber systems; however, they are not readily available for uncommon organic compounds. The solvent chosen to remove the pollutant(s) should have a high solubility for the vapor or gas, low vapor pressure, and low viscosity, and should be relatively inexpensive. Water is used to absorb VOCs having relatively high water solubilities. Amphiphilic block copolymers added to water can make hydrophobic VOCs dissolve in water. Other solvents such as hydrocarbon oils are used for VOCs that have low water solubilities, though only in industries where large volumes of these oils are available (e.g., petroleum refineries and petrochemical plants).

Another consideration in the application of absorption as a control technique is the treatment or disposal of the material removed from the absorber. In most cases, the scrubbing liquid containing the VOC is regenerated in an operation known as stripping, in which the VOC is desorbed from the absorbent liquid, typically at elevated temperatures or under vacuum. The VOC is then recovered as a liquid by a condenser.

In packed-bed scrubbers, the gas stream is forced to follow a circuitous path through the packing material, on which much of the PM impacts. The liquid on the packing material collects the PM and flows down the chamber towards the drain at the bottom of the tower. A mist eliminator (also called a "demister") is typically positioned above/after the packing and scrubbing liquid supply. Any scrubbing liquid and wetted PM entrained in the exiting gas stream will be removed by the mist eliminator and returned to drain through the packed bed. In a packed-bed scrubber, high PM concentrations can clog the bed, hence the limitation of these devices to streams with relatively low dust loadings. Plugging is a serious problem for packed-bed scrubbers because the packing is more difficult to access and clean than other scrubber designs. Mobile-bed scrubbers are available that are packed with low-density plastic spheres that are free to move within the packed bed. These scrubbers are less susceptible to plugging because of the increased movement of the packing material. In general, packed-bed scrubbers are more suitable for gas scrubbing than PM scrubbing because of the high maintenance requirements for control of PM.

Advantages of packed-bed towers include:

- Relatively low pressure drop.
- Fiberglass-reinforced plastic (FRP) construction permits operation in highly corrosive atmospheres.
- Capable of achieving relatively high mass-transfer efficiencies.
- The height or type of packing can be changed to improve mass transfer without purchasing new equipment.
- Relatively low capital cost.
- Relatively small space requirements.
- Ability to collect PM as well as gases.

Disadvantages of packed-bed towers include:

- May create water (or liquid) disposal problem.
- Waste product collected wet.
- PM may cause plugging of the bed or plates.
- When FRP construction is used, it is sensitive to temperature.
- Relatively high maintenance costs.

For gas absorption, the water or other solvent must be treated to remove the captured pollutant from the solution. The effluent from the column may be recycled into the system and used again. This is usually the case if the solvent is costly (e.g., hydrocarbon oils, caustic solutions, amphiphilic block copolymer). Initially, the recycle stream may go to a treatment system to remove the pollutants or the reaction product. Make-up solvent may then be added before the liquid stream reenters the column.

For PM applications, wet scrubbers generate waste in the form of a slurry. This creates the need for both wastewater treatment and solid waste disposal. Initially, the slurry is treated to separate the solid waste from the water. The treated water can then be reused or discharged. Once the water is removed, the remaining waste will be in the form of a solid or sludge. If the solid waste is inert and nontoxic, it can generally be landfilled. Hazardous wastes will have more stringent procedures for disposal. In some cases, the solid waste may have value and can be sold or recycled.

Configuring a control device that optimizes control of more than one pollutant often does not achieve the highest control possible for any of the pollutants controlled alone. For this reason, waste gas flows that contain multiple pollutants (e.g., PM and SO_2, or PM and inorganic gases) are generally controlled with multiple control devices, occasionally more than one type of wet scrubber.

Impingement-Plate/Tray Tower Scrubbers

This type of technology is a part of the group of air pollution controls collectively referred to as "wet scrubbers." When used to control inorganic gases, they may also be referred to as "acid gas scrubbers." When used to specifically control sulfur dioxide, the term flue-gas desulfurization (FGD) may also be used. The technology is based on the removal of air pollutants by inertial or diffusional impaction, reaction with a sorbent or reagent slurry, or absorption into liquid solvent. Applicable pollutants are primarily particulate matter, including particulate matter less than or equal to 10 μm in aerodynamic diameter (PM_{10}), particulate matter less than or equal to 2.5 μm in aerodynamic diameter ($PM_{2.5}$), and hazardous air pollutants in particulate form (PM_{HAP}); and inorganic fumes, vapors, and gases (e.g., chromic acid, hydrogen sulfide, ammonia, chlorides, fluorides, and SO_2). These types of scrubbers may also occasionally be used to control volatile organic compounds. Hydrophilic VOC may be controlled with an aqueous fluid, and hydrophobic VOC may be controlled with an amphiphilic block copolymer in the water. However, since very little data exist for this application, VOC data are not presented. When using absorption as the primary control technique, the spent solvent must be easily regenerated or disposed of in an environmentally acceptable manner.

Impingement-plate tower collection efficiencies range from 50 to 99%, depending upon the application. This type of scrubber relies almost exclusively on inertial impaction for PM collection. Therefore, collection efficiency decreases as particle size decreases. Short residence times will also lower scrubber efficiency for small particles. Collection efficiencies for small particles (<1 μm in aerodynamic diameter) are low for these scrubbers; hence, they are not recommended for fine PM control.

Control device vendors estimate that removal efficiencies range from 95 to 99%. For SO_2 control, removal efficiencies vary from 80 to greater than 99%, depending upon the type of reagent used and the plate tower design. Most current applications have an SO_2 removal efficiency greater than 90%.

The suitability of gas absorption as a pollution control method is generally dependent on the following factors: (1) availability of suitable solvent; (2) required removal efficiency; (3) pollutant concentration in the inlet vapor; (4) capacity required for handling waste gas; and (5) recovery value of the pollutant(s) or the disposal cost of the unrecoverable solvent. Impingement plate scrubbers are typically used in the food and agriculture industry, and at gray iron foundries.

FGD is used to control SO_2 emissions from coal and oil combustion from electric utilities and industrial sources. Impingement scrubbers are one wet scrubber configuration used to bring exhaust gases into contact with a sorbent designed to remove the SO_2. On occasion, wet scrubbers have been applied to SO_2 emissions from processes in the primary non-ferrous metals industries (e.g., copper, lead, and aluminum), but sulfuric acid or elemental sulfur plants are more popular control devices for controlling the high SO_2 concentrations associated with these processes.

When absorption is used for VOC control, packed towers are usually more cost effective than impingement-plate towers. However, in certain cases, the impingement-plate design is preferred over packed-tower columns either when internal cooling is desired or where low liquid flow rates would inadequately wet the packing.

Typical gas flow rates for a single impingement-plate scrubber unit are 0.47 to 35 sm³/sec (1,000 to 75,000 scfm).

Inlet gas temperature is limited to 4 to 370°C (40 to 700°F) for PM control. For gaseous pollutant control, the gas temperature typically ranges between 4 and 38°C (40 and 100°F). In general, the higher the gas temperature, the lower the absorption rate, and vice versa. Higher temperatures can lead to loss of scrubbing liquid or solvent through evaporation.

Impingement-plate scrubbers are easy to clean and maintain and are not subject to fouling as packed-bed wet scrubbers are, hence they are more suited to PM control and there are no practical limits to inlet PM concentrations.

For gas absorption applications, precoolers (e.g., spray chambers) may be needed to reduce the inlet air temperature to acceptable levels to avoid solvent evaporation or reduced absorption rates.

The following are cost ranges (expressed in third-quarter 1995 dollars) for impingement-plate wet scrubbers of conventional design under typical operating conditions, developed using EPA cost estimating spreadsheets (EPA, 1996) and referenced to the volumetric flow rate of the waste stream treated. For purposes of calculating the example cost effectiveness, the pollutant is assumed to be PM at an inlet loading of approximately 7 g/sm^3, or 3 gr/scf. The cost estimates do not include costs for post-treatment or disposal of used solvent or waste. Actual costs can be substantially higher than in the ranges shown for applications that require expensive materials, solvents, or treatment methods. As a rule, smaller units controlling a low concentration waste stream will be much more expensive (per unit volumetric flow rate) than a large unit cleaning a high pollutant load flow.

- Capital Cost: $4,500 to $25,000 per sm^3/sec ($2.10 to $11 per scfm)
- O&M Cost: $5,200 to $148,000 per sm^3/sec ($2.50 to $70 per scfm), annually
- Annualized Cost: $5,900 to $151,000 per sm^3/sec ($2.80 to $71 per scfm), annually
- Cost Effectiveness: $51 to $1,300 per metric ton ($46 to $1,200 per short ton), annualized cost per ton per year of pollutant controlled

An impingement-plate scrubber is a vertical chamber with plates mounted horizontally inside a hollow shell. Impingement-plate scrubbers operate as countercurrent PM collection devices. The scrubbing liquid flows down the tower while the gas stream flows upward. Contact between the liquid and the particle-laden gas occurs on the plates. The plates are equipped with openings that allow the gas to pass through. Some plates are perforated or slotted, while more complex plates have valve-like openings.

The simplest impingement-plate scrubber is the sieve plate, which has round perforations. In this type of scrubber, the scrubbing liquid flows over the plates and the gas flows up through the holes. The gas velocity prevents the liquid from flowing down through the perforations. Three-phase (gas-liquid-particle) contact is achieved within the froth generated by the gas passing through the liquid layer. Complex plates, such as bubble cap or baffle plates, introduce an additional means of collecting PM. The bubble caps and baffles placed above the plate perforations force the gas to turn before escaping the layer of liquid. While the gas turns to avoid the obstacles, most PM cannot and is collected by impaction on the caps or baffles. Bubble caps and the like also prevent liquid from flowing down the perforations if the gas flow is reduced.

In all types of impingement-plate scrubbers, the scrubbing liquid flows across each plate and down the inside of the tower onto the plate below. After the bottom plate, the liquid and collected PM flow out of the bottom of the tower. Impingement-plate scrubbers are usually designed to provide operator access to each tray, making them relatively easy to clean and maintain. Consequently, impingement-plate scrubbers are more suitable for PM collection than packed-bed scrubbers. Particles greater than 1 gm in aerodynamic diameter can be collected effectively by impingement-plate scrubbers, but many particles <1 μm in aerodynamic diameter will penetrate these devices.

Advantages of impingement plate scrubbers include:

- Can handle flammable and explosive dusts with little risk
- Provides gas absorption and dust collection in a single unit
- Can handle mists
- Collection efficiency can be varied
- Provides cooling for hot gases
- Corrosive gases and dusts can be neutralized
- Improves gas-slurry contact for SO_2 removal

Disadvantages of impingement plate scrubbers include:

- Effluent liquid can create water pollution problems
- Waste product collected wet
- High potential for corrosion problems
- Protection against freezing required
- Off-gas may require reheating to avoid visible (steam) plume
- Collected PM may be contaminated and may not be recyclable
- Disposal of waste sludge may be very expensive

For PM applications, wet scrubbers generate waste in the form of a slurry. This creates the need for both wastewater treatment and solid waste disposal. Initially, the slurry is treated to separate the solid waste from the water. The treated water can then be reused or discharged. Once the water is removed, the remaining waste will be in the form of a solid or sludge. If the solid waste is inert and nontoxic, it can generally be landfilled. Hazardous wastes will have more stringent procedures for disposal. In some cases, the solid waste may have value and can be sold or recycled.

For gas absorption, the water or other solvent must be treated to remove the captured pollutant from the solution. The effluent from the column may be recycled into the system and used again. This is usually the case if the solvent is costly (e.g., hydrocarbon oils, caustic solutions). Initially, the recycle stream may go to a waste treatment system to remove the pollutants or the reaction product. Make-up solvent may then be added before the liquid stream reenters the column. For FGD applications, the slurry combines with the SO_2-laden waste gas to form a waste slurry in the bottom of the scrubber. The sludge is removed from the scrubber and, depending upon the reagent or sorbent used to react with the SO_2, the waste-reacted sludge is disposed of, recycled or regenerated, or, in some cases, a salable product. For slurries that produce calcium sulfate and sulfite, oxidizing the waste sludge results in gypsum.

Gypsum is a preferred product because it can be marketed and also because of its superior dewatering characteristics. Most scrubbers are operated without the oxidizing step, and the waste sludge must be dewatered and disposed of properly. Some slurries can be regenerated and used again, but few such systems are in use due to high energy costs associated with the regeneration of the reagent.

Configuring a control device that optimizes control of more than one pollutant often does not achieve the highest control possible for any of the pollutants controlled alone. For this reason, waste gas flows that contain multiple pollutants (e.g., PM and SO_2, or PM and inorganic gases) are generally controlled with multiple control devices, occasionally more than one type of wet scrubber.

Fiber-Bed Scrubbers

This type of technology is a part of the group of air pollution controls collectively referred to as wet scrubbers. Fiber-bed scrubbers are also known as wetted-filter scrubbers and mist eliminators. The technology is based on the removal of air pollutants by inertial and diffusional interception.

Fiber-bed scrubbers are used to collect fine or soluble particulate matter or as mist eliminators to collect liquid aerosols, including inorganic (e.g., sulfuric acid mist) and volatile organic compounds. Insoluble or coarse PM will clog the fiber bed with time, and VOCs that are difficult to condense will not be collected efficiently.

Fiber-bed scrubber collection efficiencies for PM and VOC mists generally range from 70 to greater than 99%, depending upon the size of the aerosols to be collected and the design of the scrubber and the fiber beds.

Fiber-bed scrubbers are used to control aerosol emissions from chemical, plastics, asphalt, sulfuric acid, and surface coating industries. They are also used to control lubricant mist emission from rotating machinery and mists from storage tanks. Fiber-bed scrubbers are also applied downstream of other control devices to eliminate a visible plume. Despite their potential for high collection efficiency, fiber-bed scrubbers have had only limited commercial acceptance for dust collection because of their tendency to become plugged.

Fiber-bed scrubbers can treat flows from 0.5 to 47 sm³/sec (1,000 to 100,000 scfm).

The temperature of the inlet waste gas flow is generally restricted by the choice of materials. Plastic fiber beds are generally restricted to operate below 60°C (140°F).

Inlet flow loadings can range from 0.2 to 11 g/sm³ (0.1 to 5 gr/scf).

Waste gas streams are often cooled before entering fiber-bed scrubbers to condense as much of the liquid in the flow as possible and to increase the size of the existing aerosol particles through condensation. A prefilter is generally used to remove larger particles from the gas stream prior to its entering the scrubber.

The following are cost ranges (expressed in third-quarter 1995 dollars) for orifice wet scrubbers of conventional design under typical operating conditions, adapted from EPA cost estimating spreadsheets (EPA, 1996) and referenced to the volumetric flow rate of the waste stream treated. For purposes of calculating the example cost effectiveness, the pollutant is PM at a loading of approximately 7 g/sm³ (3 gr/scf) and waste gas flow ranging from 0.5 to 47 sm³/sec (1,000 to 100,000 scfm). The costs do not include costs for post-treatment or disposal of used solvent or waste.

- Capital Cost: $2,100 to $6,400 per sm³/sec ($1.00 to $3.00 per scfm)
- O&M Cost: $3,500 to $76,000 per sm³/sec ($1.60 to $36 per scfm), annually

- Annualized Cost: $4,300 to $77,000 per sm^3/sec ($2.00 to $37 per scfm), annually
- Cost Effectiveness: $40 to $710 per metric ton ($36 to $644 per short ton), annualized cost per ton per year of pollutant controlled

In fiber-bed scrubbers, moisture-laden waste gas passes through beds or mats of packing fibers, such as spun glass, fiberglass, or steel. If only mists are to be collected, small fibers may be used, but if solid particles are present, the use of fiber-bed scrubbers is limited by the tendency of the beds to plug. For PM collection, the fiber mats must be composed of coarse fibers and have a high void fraction, to minimize the tendency to plug. The fiber mats are often sprayed with the scrubbing liquid so particles can be collected by deposition on droplets and fibers. For PM removal, the scrubber design may include several fiber mats and an impingement device. The final fiber mat is typically dry for the removal of any droplets that are still entrained in the gas stream.

Advantages of fiber-bed scrubbers include:

- Can handle flammable and explosive dusts with little risk.
- Can handle mists.
- Relatively low pressure drop.
- Provides cooling for hot gases.
- Corrosive gases and dusts can be neutralized.

Disadvantages of fiber-bed scrubbers include:

- Effluent liquid can create water pollution problems.
- Waste product collected wet.
- High potential for corrosion problems.
- Protection against freezing required.
- Collected PM may be contaminated and may not be recyclable.
- Disposal of waste sludge may be very expensive.

For liquid aerosols, the used scrubbing liquid must be treated to remove the captured pollutant from the solution. The effluent from the column may be recycled into the system and used again. This is usually the case if the solvent is costly (e.g., hydrocarbon oils, caustic solutions). Initially, the recycle stream may go to a waste treatment system to remove the pollutants or the reaction product. Make-up scrubbing liquid may then be added before the liquid stream reenters the column.

For PM applications, wet scrubbers generate waste in the form of a slurry. This creates the need for both wastewater treatment and solid waste disposal. Initially, the slurry is treated to separate the solid waste from the water. The treated water can then be reused or discharged. Once the water is removed, the remaining waste will be in the form of a solid or sludge. If the solid waste is inert and nontoxic, it can generally be landfilled. Hazardous wastes will have more stringent procedures for disposal. In some cases, the solid waste may have value and can be sold or recycled.

ACTIVATED CARBON AND OTHER ADSORBER SYSTEMS

Adsorption is defined as the collection and concentration of a substance (e.g., an organic solvent) onto a solid surface (e.g., carbon adsorbent) from a gas (or liquid). The solvent is bound to the carbon due to the attractive forces (affinity) between the carbon and the organic solvent.

Activated carbon adsorption is a popular method of recovering solvents from air emissions as well as wastewater. At one time, adsorption systems were largely viewed as an end-of-pipe pollution control technology as opposed to today's view of a solvent recovery system. This view has altered because of technology advances that have enhanced the ability of adsorbers to recover (desorb) the solvent from the activated carbon. Further advancements in the process of regenerating the used activated carbon have made the need for replacing the adsorbent less frequently than in past practices, thus making this an economical technology.

Adsorption traditionally has found wide application in the field of odor control. Odors are generally created by the presence of small quantities of contaminants in large volumes of air. The equilibrium relationships associated with adsorption, unlike those involved in absorption, lend themselves to removal of low-concentration contaminants. However, because of the complex mass transfer mechanisms involved in adsorptions, and the great variability in adsorbent physical properties that strongly affect performance, adsorption is much less amenable to generalized design from basic physical data than is absorption. Good design requires specific data for the gas-solid system involved.

In industrial operations, adsorption is accomplished primarily on the surfaces of internal passages within small porous particles. Three basic mass transfer processes occur in series: (1) mass transfer from the bulk gas to the particle surface, (2) diffusion through the passages within the particle, and (3) adsorption on the internal particle surfaces. Each of the processes depends on the system operating conditions and the physical and chemical characteristics of the gas stream and the solid adsorbent. Often, one of the transfer processes will be significantly slower than the other two and will control the overall transfer rate. The other process will operate nearly at equilibrium.

Heat transfer may also play an important role in an adsorption system. The adsorption process is exothermic. Physical adsorption equilibria behave in a manner similar to vapor-liquid equilibria; transfer out of the vapor phase is favored by decreasing temperature. Therefore, rapid dissipation of heat away from the adsorbing surface improves adsorption performance. Chemisorption rates, on the other hand, generally increase with increased temperature.

External mass transfer is the only process of the three involved in adsorption that can be predicted with reasonable accuracy from physical data. Mass transfer from the bulk gas to the particle surface can be considered by the film resistance approach. The rate of mass transfer is proportional to the external surface area of the adsorbent particles and the adsorbate concentration difference between the bulk gas and the particle external surface. The proportionality constant is the mass transfer coefficient, the reciprocal of the resistance to mass transfer of a hypothetical thin film at the particle surface. Correlations of mass transfer coefficients as functions of particle size, gas flows, and properties and operating conditions have been summarized extensively in the literature. The references cited at the end of this chapter will assist you.

Transfer of material from the particle surface to internal adsorption sites is accomplished by diffusion through the internal passages of the particle. Diffusion takes place by one or more of several mechanisms. Where passages are sufficiently large that intermolecular collisions are more likely than collisions with the passage walls, bulk diffusion predominates. In smaller passages, and at lower pressures, where collisions with the passage walls are more probable, Knudson diffusion controls. Knudson diffusion rates are often an order of magnitude lower than bulk rates for a given pore length. A third diffusion mechanism, surface diffusion, may predominate in small-passage, high-surface situations. Surface diffusion is the migration of molecules along the passage surface after adsorption. Because of the irregularity and variability of intraparticle passages, and the difficulty of measuring surface concentrations, intraparticle mass transfer is extremely difficult to predict accurately from physical data. Its effects can be determined to a degree by comparison of experimental adsorption rates for different size adsorbent particles, allowing for differences in external mass transfer rates.

The actual adsorption of vapor molecules takes place mainly on the surface of internal passages within the adsorbent particles, since that is where most of the available surface exists. The adsorption process may be either physical or chemical in nature. Physical adsorption is a readily reversible process that occurs as a result of the physical attraction between the gas molecules and the molecules of the solid surface. If the gas-solid intermolecular attraction is greater than the intermolecular attractions in the gas phase, the gas will condense on the solid surface, even though its pressure is lower than its vapor pressure at the prevailing temperature. For example, the equilibrium adsorption pressure of acetone on activated carbon may, under some conditions, be as little as 150 to 1,100 of the equilibrium vapor pressure at the same temperature. Therefore, small concentrations of contaminants can often be removed from gas streams by adsorption, but not by absorption. The heat released on adsorption is usually somewhat greater than the latent heat of vaporization and of the order of the heat of sublimation of the gas. Chemical adsorption, or chemisorption, involves formation of chemical bonds between the gas and surface molecules.

The strength of the chemical bond may vary considerably, and identifiable chemical compounds in the usual sense may not actually form. However, the adhesive force is much greater than found in physical adsorption. The heat liberated during chemisorption is usually large, of the order of the heat of chemical reaction.

Most of the adsorption data available from the literature are equilibrium data. Equilibrium data are useful in determining the maximum adsorbent loading that can be obtained for a specific adsorbate-adsorbent system under given operating conditions. However, equilibrium data by themselves are insufficient for design of an adsorption system. Overall mass transfer rate data are also necessary.

The equilibrium adsorption characteristics of gas or vapor on a solid resemble in many ways the equilibrium solubility of a gas in a liquid. Adsorption equilibrium data are usually portrayed by isotherms: lines of constant temperature on a plot of adsorbate equilibrium partial pressure versus adsorbent loading in mass of adsorbate per mass of adsorbent. Isotherms take many shapes, including concave upward and downward, and S-curves. Equilibrium data for a given adsorbate-adsorbent system cannot generally be extrapolated to other systems with any degree of accuracy.

Several useful methods are available for extrapolating equilibrium data for a given system to various temperatures and pressures. One convenient method is by use of a reference substance plot. Here, the adsorption equilibrium partial pressure of the adsorbate is plotted against a pure substance vapor pressure, preferably that of the adsorbate. If logarithmic coordinates are used on both axes, lines of constant adsorbent loading, isosteres, are linear for most substances. Therefore, only two datum points are required to establish each isostere.

Isosteres will also generally be linear on a plot of the logarithm of the ratio of pure substance vapor pressure to adsorption equilibrium vapor pressure against reciprocal absolute temperature. A further condensation of data may be accomplished by plotting absolute temperature times the logarithm of the ratio of pure substance vapor pressure to adsorption equilibrium vapor pressure at that temperature against adsorbent loading. For most singly adsorbate systems, a single curve will result for all temperatures, at least over a moderate temperature range.

Mass Transfer Rate Considerations. As discussed previously, the mass transfer mechanism involved in industrial adsorption processes is complex. Generally, basic physical data on the materials involved are insufficient for design. Experimental mass transfer rate data for the specific adsorbate-adsorbent system are usually required for good design.

Three basic schemes are used in adsorption systems for obtaining effective gas-solid contact: (1) the fluidized bed, (2) continuous moving bed, and (3) unsteady-state fixed-bed techniques. By far the most common is the unsteady-state fixed-bed adsorber. In this system, the contaminated gas is passed through a stationary bed of adsorbent. The bed is operated in this manner until the contaminant level in the effluent begins to rise. The adsorbent must then be replaced or regenerated. Generally gas flow is diverted to a second, parallel bed to allow continued operations during adsorbent change or regeneration. Adsorbent beds range in size and form from small disposable cartridges to dumped beds contained in large vessels. Unsteady-state fixed-bed adsorbers have the advantage of being relatively simple and economical, particularly at low adsorbate rates. Since the bed is stationary, the adsorbent is handled only during replacement, which should be infrequent in a well-designed system. Continuous solids-handling systems and their inherent high cost and mechanical problems are avoided.

The primary disadvantage of fixed-bed adsorbers arises when contaminant rates are high. Because of the unsteady-state nature of the operation, a large portion of the in-process adsorbent inventory is saturated and, therefore, inactive.

Where adsorbate rates are high, unduly large beds are required. In addition, gas flow rates through fixed beds are limited by pressure drop. Extremely high gas rates may require uneconomically large beds.

In fluidized-bed adsorbers, the combination of high gas rate and small adsorbent particle size results in suspension of the adsorbent, giving it many of the characteristics of a fluid. Fluidized-bed adsorbers, therefore, lend themselves to truly continuous, countercurrent, multistage operation. Adsorbent inventory is minimized.

Fluidized-bed adsorbers have several disadvantages. The continuous handling and transport of solids is expensive from an equipment standpoint; fluidized-bed systems must be large to be economical. Solids handling also presents a potential for mechanical problems. Careful control is required to keep the adsorbent fluidized, while minimizing adsorbent loss with the gas-phase attrition of the adsorbent can be high, requiring substantial makeup.

Continuous moving-bed adsorbers, like fluidized-bed systems, lend themselves to true countercurrent, multistage operation. The adsorbent, however, is not fluidized, but is mechanically converged or falls by gravity through the rising stream of gas. Attrition is generally higher than in fluidized-bed systems, but control may be less critical. Other advantages and disadvantages of fluidized-bed adsorbers apply to moving-bed adsorbers.

When adsorbate rates are sufficiently high to make periodic adsorbent replacement uneconomical, regeneration of the adsorbent can usually be justified. In continuous, steady-state systems, regeneration is required for economical operation.

Adsorbent regeneration is normally accomplished by reversing the adsorption process, either by decreasing the system pressure or, more commonly, by increasing the system temperature. In some cases, particularly in chemisorption systems, the adsorbent activity can be restored by reaction with a suitable reagent.

One of the inherent problems in regeneration of adsorbent beds is disposal of the desorbed material. In activated carbon systems, the most common in odor control applications, regeneration is accomplished by heating the bed with a gas or vapor that carries the desorbed contaminant out of the bed. The most commonly used carrier is superheated steam. Normally, the adsorbate is condensed along with the steam. If large quantities of adsorbate are involved, or if the adsorbate is highly water soluble, a secondary liquid waste disposal problem may result. The energy costs for a superheated steam-condensing regeneration system can be appreciable for a large adsorption system. A pollution prevention practice is to use hot combustion gases as the regeneration carrier. The energy required to heat the gas is supplied by combustion of the desorbed material; supplemental fuel is supplied, if necessary. Both the energy cost and secondary disposal problems are solved with this system. However, the cost of the additional equipment required makes this system uneconomical for small adsorption installations.

There are several choices for the adsorbent. Activated carbon still remains the most widely used, especially for VOCs. Activated carbon is by far the most commonly used adsorbent in odor control applications and many VOC recovery applications. Because of its relatively uniform distribution of surface electrical charge, activated carbon is not selective toward polar molecules.

It can, therefore, be used to remove many organic vapors from gas streams with high water vapor contents. Water molecules, being highly polar, show strong attractions for each other, which compete with their attractions for the nonpolar carbon surface. Consequently, the large, less polar organic molecules are selectively adsorbed. Activated carbon is most effective for adsorbing organic materials that boil at normal ambient temperature or higher. In general, effectiveness increases with increasing molecular weight. Activated carbons have surface areas on the order of 300 to 700 ft^2/ft^3. Average pore diameter ranges from 20 to 40 microns, typical for most commercially used adsorbents. However, the distribution of pore sizes is substantially broader than found in other adsorbents.

The most commonly used adsorbents of the siliceous class are silica gels and synthetic zeolites, or molecular sieves. These materials are available over a wide range of adsorbent capacities. At best, their capacities are of the same order of magnitude as that of the most highly activated carbons. They exhibit a greater selectivity for polar molecules than does activated carbon. Silica gel is commonly used to remove water from gas streams. Surface areas average 200 to 700 ft^2/ft^3. Average pore diameters of various grades range from 20 to 140 microns.

Molecular sieves are synthetic zeolites that can be manufactured with extremely close control of pore size. Therefore, they can be tailored to suit specific applications. In addition to gas drying applications, molecular sieves are used for the separation of gases and vapors on the basis of molecular size and shape. Surface areas range from 350 to 1000 ft^2/ft^3.

Since metals are less electrophilic than silicon, metal oxide adsorbents show even stronger selectivity for polar molecules than do siliceous materials. The most commonly used metal oxide adsorbent is activated alumina, used primarily for gas drying. Occasionally, metal oxides find applications in specific chemisorption systems. For example, several processes are under development utilizing lime or limestone for removal of sulfur oxides from flue gases. Activated aluminas have surface areas in the range of 200 to 1000 ft^2/ft^3. Average pore diameters range from about 30 to 80 microns.

For some applications, an adsorbent may be impregnated with a material that enhances its contaminant-removal ability. The improved effectiveness may be related to any of several mechanisms. The impregnating material may react with the vapor contaminant to form a compound or complex that remains on the adsorbent surface. Some impregnants react with the contaminant, or catalyze reactions of the contaminant with other gas constituents, to form less noxious vapor-phase substances. In some instances, the impregnant acts as a catalyst intermittently, for example, under regeneration conditions. In this case, the contaminant is adsorbed by physical adsorption and destroyed by a catalytic reaction during regeneration.

THERMAL DESTRUCTIVE TECHNOLOGIES

For more than a decade now, environmental authorities have targeted the baking industry, chemical process industry, pharmaceutical industry, fabric coating, rubber extrusion, electronic components, soil remediation, metal coating, wood working, formaldehyde, and sterilizers. The reason for this is clear: These industry sectors are large generators of VOC emissions. Table 5.1 provides a list of industry operations and the typical VOC, solvents, and other off-gases associated with them.

The solutions that are available to control these emissions are normally dictated by the volume of air that is to be

TABLE 5.1   Examples of Gaseous Pollutants by Industry

Industry Operation	Gaseous Pollutants Emitted
Acetate finish coating	Silicate solution
Alcohol synthesis	C_1, C_2, C_3, C_6 hydrocarbons
Automobile coating	Ketones, xylene, toluene, phenols
Bakery ovens	Ethanol
Can coating	Ketones, alcohols, aromatic hydrocarbons, ethers
Chemical processing	All of the above
Coffee roasting	Heavy oils from coffee beans
Coil coating	Phosphates, solvesso, cyclohexanol, alcohols, carbitols, hydrocarbons
Electronic components	Butyl acetate, xylene, MEK, cellosolve
Fiberglass coating	Teflon emulsion, fiberglass, synthetics, styrene
Flexographic printing	Flexographic ink derivatives, alcohols, glycol
Formaldehyde	Formaldehyde, methanol, CO
Lithographic print/paint	Butyl cellosolve, ciacetone alcohol, solvesso, cellosolve acetate, xylene, MIBK
Metal coating	Alcohols, cellosolve acetate, phthalates, solvesso
Paper coating	High-boiling organics, latexes
Pharmaceuticals	Isopropanol, toluene, hydrocarbons
Phthalic anhydride mfg.	Organic acids
Resin plant	Formaldehyde, phenols, phthalic anhydride
Rubber Processing	
Soil remediation	Benzene, toluene, ethylene, xylene
Sterilizers	Ethylene oxide
Vinyl surgical glove	Polyvinyl chloride, cioctyl phthalate
Wire enameling	Cellosolve acetate

processed. The volume of air flow, measured in cubic feet per minute, is designated as Actual Cubic Feet per Minute (ACFM) or SCFM, where "S" stands for "standard" cubic feet per minute, at 70°F, sea level, and 1 atmosphere.

Thermal destructive techniques have been widely used for many years to control some of these emissions. Thermal oxidizer sizes range from 100 scfm up to 100,000 scfm. Each industry has operations that dictate the exhaust flow that must be processed.

Commercially available thermal oxidizer systems are pre-engineered; that is, the equipment is designed on the principle that in order for the equipment to be competitive in the marketplace, then a series of products of fundamentally standard designs are tailored to the application by changing some of the parameters as dictated by the requirements. This is not always the case with other pollution control systems, as oftentimes custom-built systems are specified. Since thermal oxidation equipment has a burner, the designs require controls for safety and operation.

Controls are termed user friendly with control logic being relay or computer based. As noted in earlier chapters, particular hydrocarbons affect health in several ways. Some affect the respiratory system, while others are air toxins. Recall from Chapter 1 that the EPA issued new rules and regulations in 1990 regarding VOCs and air toxins. Subsequent controls or regulations have been issued that have identified air toxins as particularly dangerous to industrial workers. One of the major air toxins is benzene, and benzene is one of the principal components of gasoline. The remediation of benzene from underground storage tanks and leaky underground petroleum bases is a significant source of air toxins. Carbon tetrachlorethylene, hydrogen cyanide, and ketones are all air toxins being controlled or regulated, since they are carcinogenic and may create an immediate problem.

In determining the most appropriate technology to controlling a process stream, it is necessary to characterize the airstream. The first step in the characterization is to establish what organics and other contaminants need to be destroyed or controlled. Recognize also that the sizing of the equipment required is dependent on the air flow and the organic loading.

There are four general oxidation technologies:

- Catalytic recuperative
- Thermal recuperative
- Regenerative thermal, regenerative catalytic
- Rotor concentrator

Low-flow, low-concentration streams are best handled by a catalytic recuperative oxidizer. When the concentration of the stream is between 15 and 20% Lower Explosion Limit (LEL) then either a catalytic recuperative or thermal recuperative is the best technology. For process streams between 20 and 25% LEL then thermal recuperative is the preferred solution.

It is important to establish the volume of the process stream that is to be treated. There are companies that have not needed to buy pollution control equipment in the past. Their approach to eliminating pollution has been the more exhaust, the better. This is particularly true in the low margin industries. The more complex and high-volume industries, e.g., the chemical industry, have normally been concerned about the air exhausted from their plant and have paid attention to the exhaust volume by process control. The industries that have not paid attention to exhaust volume are not sufficiently aware that when buying pollution control equipment the cost of that equipment is going to depend on two elements:

- The air volume that must be processed
- The pounds per hour of organics that must be processed

Process flow reduction should be targeted and must be in such a way that it will not impact upon the operation of the process or plant. The normal first response by a plant engineer is that air flow reduction cannot be done. However, subsequent investigation has resulted in that most of the time flow reduction can be done. This should be a significant area to focus on when conducting a pollution prevention audit.

Determining the concentration of the organic matter in plant off-gases can be difficult. However, in an operation where solvents or other organics are purchased, the number of pounds or tons of the organics that have been bought in the past year is known. Note that this information must also be supplied to the environmental authorities. Hence, examining purchasing records can provide some good insight into how much is ultimately lost as emissions. Given the tons of organics purchased in the previous years, and the number of operating hours per year, the ratio of those two numbers yields the pounds per hour, a conservative initial estimate of the average organic loading. This methodology assumes that everything that is purchased will be emitted up the stacks. In addition to the loading, it is necessary to determine if there are compounds in the exhaust stream that can be deleterious to catalysts used, thereby precluding the use of this technology.

As an example, when automotive catalytic mufflers and converters were introduced many years ago, the automobile industry required the petrochemical industry to eliminate lead from gasoline since lead degraded and reduced the effectiveness of the catalyst and caused the destruction of the gasoline. One set of industrial compounds that can harm catalysts are halogens, a family of compounds that include chlorine, bromine, iodine, and fluorine. Bromine, while not prevalent in industry, is present in chemical plants. Freons are fluorine compounds. Silicone is another compound that is deleterious to catalysts. It is used as a slip agent, or a lubricant, in many industrial processes. Phosphorous, heavy metals (zinc, lead), sulfur compounds, and any particulate can result in shortening the life of the catalyst. It is necessary to estimate the volume or the amount of each of those contaminants, to assess the viability of catalytic technologies for the application.

RTO/RCO and Rotor/Concentrator systems are typically higher in capital costs, but the operating cost savings on high-volume, low-concentration streams make these technologies attractive. Subsequently the capital and operating cost of the equipment and the cost of the installation of that equipment is related to the sizing. Since oxidation equipment requires fuel and electricity, the operating costs will also depend on the hours or operation per day, per week, and per year. In addition, the costs depend on the quantity of organics that need to be processed, in pounds per hour (lb/hr), or in parts per millions by volume. A critical consideration is whether the organic loading is steady state or if it has peaks or spikes. Since all oxidation equipment has to operate at an elevated temperature, it is necessary to know the type of auxiliary energy of fuel available. All oxidizers can operate on electric heating, natural gas, or LPG, whereas thermal oxidizer can operate with sulfur-based fuels such as No. 2 or No. 6 fuel oil. Electric heat is only suitable for small air flows; otherwise it is too costly. No thermal oxidizer should be considered without some arrangement for heat recovery, because a continuing cost of air pollution control is the operating cost. Available heat recovery efficiencies vary depending on the type of thermal oxidation technology selected. The specific heat recovery efficiency selected for an application depends on the organic concentration. Table 5.2 shows an economic comparison for a particular airstream.

The comparison is for a process stream of 10,000 scfm with an inlet temperature of 70°F. Toluene is the contaminant in this example. Toluene has a calorific content of 16,720 btu/lb and a LEL of 1% to 10% by volume. For each of the technologies, an economic analysis was performed at a toluene loading of 1% LEL and at 10% LEL. The results of this analysis are reported as operational costs in $/hr. The operational cost is the sum of the fuel usage as well as the electricity needed to run the system fans. All the technologies are assumed to be running on natural gas at a cost of $4.50/MMBtu. The electricity price has been taken at $0.06/kWh. A catalytic recuperative oxidizer, with a 65% effective heat exchanger, is more economical to operate at the 1% to 10% LEL range than a thermal recuperative oxidizer with a 65% effective heat exchanger. The thermal recuperative oxidizer will only begin to show a significant reduction in operational costs around the 15 to 25% LEL range. Both of the regenerative oxidizers utilize a 95% effective heat exchanger. These technologies are best suited for high air flow, low loading process streams.

TABLE 5.2   Operating Cost Comparisons for Different Systems

Process Stream% LEL	Catalytic Recuperative Oxidizer	Thermal Recuperative Oxidizer	Regenerative Catalytic Oxidizer	Regenerative Thermal Oxidizer	Rotor Concentrator with Thermal Oxidizer
1	$11.00/hr	$28.37/hr	$2.50/hr	$5.65/hr	$0.30/hr
10	$3.70/hr	$16.72/hr	$1.65/hr	$1.60/hr*	N/A

Operational Modes: The Catalytic Recuperative Oxidizer assumes a 65% efficienct heat exchanger; the Thermal Recuperative Oxidizer assumes a 65% efficienct heat exchanger; the Regenerative Catalytic Oxidizer assumes a 95% efficienct heat exchanger; the Regenerative Thermal Oxidizer assumes a 95% efficient heat exchanger; the rotor concentrator wheel assumes a 6:1 concentration ratio.

Both of the regenerative oxidizers begin to show high-temperature bypass at the 10% LEL loading. This means that the loading point at which the oxidizers would require zero supplemental fuel has been exceeded. Excess heat is now being produced in the oxidizer. This heat must be able to escape from the oxidizer by way of a high-temperature bypass. Also, the rotor concentrator wheel is another technology suited for high airflow, low loading process streams. The rotor concentrator will concentrate the process stream from anywhere between a 6:1 to a 13:1 concentration ratio. Thus, the airflow will be reduced 6 times, while the contaminant loading will be increased 6 times. The rotor concentrator technology is not suited for process streams at a 10% LEL loading.

In the U.S. and many parts of Western Europe, 95% destruction of VOC is required, and when toxic chemicals are present the destruction efficiency is raised to 99%. Characterization of the VOC-laden stream assists in selecting the appropriate technology to achieve the required destruction efficiency.

There are seven fundamental oxidizer technologies that achieve the oxidation of organics in alternate ways. These technologies are:

- Catalytic recuperative
- Thermal recuperative
- Catalytic regenerative
- Thermal regenerative
- Rotor concentrators
- Flare/burner
- Ceramic filter

Before describing these, some general background information is presented.

Thermal oxidation, by definition, converts a hydrocarbon, in the presence of oxygen and heat, to carbon dioxide and water vapor. A general equation showing this relationship is:

$$C_nH_{2m} + (n + m/2)O_2 \rightarrow nCO_2 + mH_2O + \text{Heat}$$

The particular n and m subscripts of the equation are used to define the number of carbon atoms and hydrogen atoms. The number of oxygen atoms convert to n molecules of carbon dioxide and m molecules of water vapor and heat, which is given off in the exothermic reactor.

One measure of the organic loading or the organic concentration of the process stream is the "lower explosive limit," (also known as the "lower flammable limit," designated as the LFL). The LEL of a stream is the lowest organic concentration in a stream that would, in the presence of an ignition source, yield a combustible mixture. At this concentration, the stream is said to be at 100% LEL. As examples, if a steam has toluene or alcohol and it is at 100% LEL, this means that the concentration within that stream is explosive if a spark or ignition source is present (recall the so-called fire-triangle explanation of combustion). Look back in Chapter 3 for some common organics and their fire properties.

Let's take the example of benzene, which at 12,000 ppm is 100% LEL. The National Fire Protection Association (NFPA) states that equipment can operate, without LEL monitors or controls, if the LEL is less than 25% LEL. For benzene, then, 25% LEL is equal to 3000 ppm. This upper boundary becomes a dictating factor in the selection and design of the oxidation equipment. If the concentration is higher than 25% LEL, the NFPA requirements state that an LEL monitor is required. Using an LEL monitor, NFPA guidelines allow operation up to 50% LEL (a 2:1 safety factor). Thus, 100% LEL is explosive; if the stream is at 25%, a factor safety of 4 exists.

In addition to the explosive aspects of the LEL, another issue is the heat energy given off during oxidation. An estimate of the exotherm is that there will be a 25°F rise per 1% LEL in the stream. Hence, if the process air enters the oxidizer at a given temperature, and if the stream has a concentration of 2% LEL, then a 50°F rise in process stream temperature is expected after oxidation. If the process stream were running at a 10% LEL, then a 250°F temperature rise would be predicted. A maximum LEL of 25% yields a 625°F temperature rise of the process stream.

The concentration that is required to yield 100% LEL varies with the characteristics of the identified organics. For example, ethanol has a 20,000 ppm by volume concentration, whereas benzene has a 12,000 ppmv.

Not only do different organics have different LELs but every time a pound of a different organic is oxidized, a different amount of heat will be given off. An example that demonstrates this energy release is the catalytic muffler in an automobile. When mufflers were initially installed, the high level of unburned gasoline that went through the muffler caused excessive heating and subsequently caused fires if the auto had been parked on leaves. The reason for the excessive heat was that for each pound of gasoline being oxidized 20,000 Btu's were being released.

Chlorinated organics are hydrocarbons that have one or many chlorine atoms. Oxidation of chlorinated hydrocarbons yields CO_2, water vapor, and hydrogen chloride (HCl) gas. Some typical chlorinated organics are TCE and PCE. These organics have calorific values as low as 5000 BTU/lb.

We should not forget to mention the importance of the catalyst. The characteristics of oxidation catalyst are many and varied. Fundamentally, if an airstream containing organics is heated and passed across a catalyst, the organics will be converted to carbon dioxide and water vapor. However, the percentage conversion happens at different temperatures for different organics and for different catalysts. As an example, to destroy 25% of the toluene in an airstream, a temperature of about 300°F is required. However, the EPA does not require 25% destruction, but 95 to 99% destruction. In order to destroy 95% of the toluene, 500 to 550°F is required. Some organics require higher temperatures to be destroyed than others, catalytically. Alcohols, isopropyl alcohol, and ethanol can be destroyed relatively simply, whereas the acetates, particularly the ethyl acetates and propyl acetates, may require temperatures in the 750°F range in order to achieve adequate destruction. Depending on the process stream, either a single organic may be present, as found in the chemical industry, or a multiplicity of organics exists, as in printing operations. Having a multiplicity of organics imposes the requirement of focusing on the ability to destroy the most difficult organic constituent.

Some organics cannot be effectively destroyed by catalyst. For example, heptane and hexane can be destroyed at temperatures of 600 to 700°F, whereas propane, ethane, and methane require temperatures beyond a reasonable temperature range. Since methane is not a smog-producing organic, a guarantee to destroy 95% of the organics means that the methane is not considered and is removed from that stream in evaluating the process. However, if propane is the auxiliary fuel, that is, if the burners are being driven by LPG or by propane directly, it means that there will be contribution to the VOC at the end of the stack.

The catalyst is normally contained on a ceramic substrate. These ceramics are extruded in a malleable state and then fired in ovens. The process consists of starting with a ceramic and depositing an aluminum oxide coating. The aluminum oxide makes the ceramic, which is fairly smooth, have a number of bumps. On those bumps a noble metal catalyst, such as platinum, palladium, or rubidium, is deposited. The active site, wherever the noble metal is deposited, is where the conversion will actually take place. An alternate to the ceramic substrate is a metallic substrate. In this process, the aluminum oxide is deposited on the metallic substrate to give the wavy contour. The precious metal is then deposited onto the aluminum oxide. Both forms of catalyst are called monoliths.

An alternate form of catalyst is pellets. The pellets are available in various diameters or extruded forms. The pellets can have an aluminum oxide coating with a noble metal deposited as the catalyst. The beads are placed in a tray or bed and have a depth of anywhere from 6 to 10 in. The larger the bead (1/4 in versus 1/8 in), the less the pressure drop through the catalyst bed. However, the larger the bead, the less surface area is present in the same volume, which translates to less destruction efficiency. Higher pressure drop translates into higher horsepower required for the oxidation system. The noble metal monoliths have a relatively low pressure drop and are typically more expensive than the pellets for the same application.

An alternate to a noble metal catalyst is a base metal catalyst. A base metal catalyst can be deposited on a monolithic substrate or is available as a pellet. These pellets are normally extruded and hence are 100% catalyst rather than deposition on a substrate. A benefit of base metal extruded catalyst is that if any poisons are present in the process stream, a deposition of the poisons on the surface of the catalyst occurs. Depending on the type of contaminant, it can frequently be washed away with water. When it is washed, abraded, or atritted, the outer surface is removed, and subsequently a new catalyst surface is exposed. Hence, the catalyst can be regenerated. Noble metal catalyst can also be regenerated but the process is more expensive. A noble metal catalyst, depending on the operation, will typically last 30,000 hours. As a rule of thumb, a single shift operation of 40 hours a week, 50 weeks a year, results in a total of 2000 hours per year. Hence, the catalyst might have a 15-year life expectancy. From a cost factor, a typical rule of thumb is that a catalyst might be 10 to 15% of the overall capital cost of the equipment.

A catalytic recuperative oxidizer consists of several main elements:

- System fan
- Heat exchanger
- Reactor
- Catalyst
- Exhaust stack

With this technology, the process stream is ducted to the oxidizer and enters a system fan. The system fan is driven by a motor, and the process stream is forced under positive pressure through a heat exchanger. The heat exchanger is usually a cross-flow heat exchanger of the shell and tube type. Plate-type heat exchangers can also be used in the application. Normally, the process stream enters through the tube side of the heat exchanger, to clean the tubes. As the air enters in and goes through the heat exchanger, it is heated and will then exit into the reactor. As it enters the reactor, the process stream will be further heated by a burner, controlled by a thermocouple measuring the temperature of the air and a temperature controller regulating the burner firing, to bring the process stream up to the catalyzing temperature of 300 to

700° F. The catalyzing temperature depends on the organic, the requirement for the destruction of the organic and the type and volume of catalyst. At the catalyzing temperature, the process stream will pass through a series of beds, having catalyst in them. As the air containing organics comes across the catalyst, the organic is converted to CO_2 and water vapor and an exothermic reaction occurs. This exotherm will raise the temperature of the stream exiting the catalyst bed. Hence, the catalyst outlet temperature will be higher than the temperature going into the catalyst bed. The process stream is then directed through the shell side of the heat exchanger, where it preheats the incoming air and is then exhausted to the atmosphere.

Thermal Recuperative Oxidizer. The best way to understand the theory of operation for thermal recuperative oxidation is by the three "Ts" of combustion (time, temperature, and turbulence). Time refers to the retention time or residence, which is the length of time that an organic is at the appropriate oxidation temperature. Roughly, if a 95% destruction efficiency is required, a residence time of a half a second is adequate. That is, the organic compound is brought up to a temperature of about 1400°F and maintained at that temperature for a retention time of half a second. Both retention time and turbulence must be achieved, where turbulence implies a required degree of mixing. If the turbulence occurs and if the 1400°F for a half a second is achieved, with adequate oxygen, then 95% of the organic will be destroyed. In some cases, oxygen must be added to a process stream. For each organic, a specific temperature and a specific retention time is required to achieve the desired level of destruction. One of the concerns that exists in thermally processing chlorinated organics is that the by-products of combustion that are formed may be more harmful than the initial product that is to be destroyed. A by-product of combustion is a compound formed, either catalytically or thermally, when most of the organic compound is converted to CO_2 and H_2O. For example, when 97% of the toluene in an airstream is destroyed, what happens to the other 3%? Does it remain as toluene or are new products formed? Frequently, "by-products of combustion" are formed, and it is not possible to predict what those by-products are. They could be a mixture of the original compounds, or they could be more harmful. The concern that we have in processing chlorinated organics thermally at these high temperatures is that dioxins may be generated. Dioxins are in effect much more harmful than the organic that will cause VOCs or smog in the air.

Thermal oxidizers must be built to provide the residence time and temperatures to achieve the desired destruction efficiency (DE). As such, thermal oxidizers are comparatively larger than catalytic oxidizers, since their residence time is two to four times greater. Historical designs of thermal oxidizers were comprised of carbon steel for the outer shell and castable refractory or brick as the thermal liner (a refractory is like a cement, which is put on the inside of the reactor shell to act as a thermal insulation barrier). Modern units are designed and built using ceramic fiber insulation on the inside, which is a lightweight material and has a relatively long life. Old refractory would tend to fail over a period of years by attrition of expansion and contraction.

A forced draft thermal recuperative oxidizer consists of a fan that forces the process airstream through a heat exchanger. As the process flow exits the heat exchanger the burner fires and the process stream is brought up to temperature. The reactor chamber, which is lined with a high-temperature ceramic fiber, is designed for the required retention time. A shell and tube heat exchanger is normally used for thermal oxidizers because of expansion concerns. The material selection for the shell and tube heat exchanger needs to be a high-grade stainless, adequate for expected steady-state and peak temperatures.

In addition to forced draft systems, induced draft systems are also used. The induced draft system is slightly more expensive but is recommended when particulate or organic oils are present. Particulates impact upon a forced draft fan and will have a negative effect on the system performance. If the process stream is clean, a forced draft system is appropriate.

The burners used are modulating burners; that is, if the organic is preheated to an adequate temperature, the burners will modulate down to 0 so that there is no energy required for the continued oxidation.

One of the advantages of the thermal recuperative oxidizer is that it is possible to process organics that may be a poison or be detrimental to the catalyst. In addition, if the organic concentration is very high, for example, the organic level is of the 20 to 25% LEL, then thermal recuperative oxidation is appropriate.

For relatively large-volume and low-concentration streams the disadvantage with the thermal recuperative design is that the metallic heat exchanger only recovers about 70% of the heat, and operating costs increase dramatically. The level of heat exchanger efficiency is limited by material cost to increase heat exchanger efficiency and the fact that autoignition, where organics ignite due to temperature alone, becomes a problem with the life of metallic heat exchangers.

Regenerative Thermal Oxidizer (RTO). The Regenerative Thermal Oxidizer is a thermal oxidizer consisting of two or more ceramic heat transfer beds, which act as heat exchangers, and a purification chamber or retention chamber, where the organics are oxidized and converted to CO_2 and H_2O vapor. The operation of an RTO Dual-Bed requires the initial preheating of the ceramic heat transfer beds to a temperature of 1500°F during the start-up mode. This is accomplished by operation of fuel-fired burner located in the purification chamber. To equalize the preheating of the ceramic heat transfer beds, the air is directed into and out of the ceramic heat transfer beds by operation of pneumatic diverter valves, located under each ceramic heat transfer bed. During initial start-up, outside air is supplied to the oxidizer through the make-up air damper tee located on the inlet side of the process air fan. A Programmable Logic Controller (PLC) monitors and controls the direction of the airflow.

After the ceramic heat transfer beds have reached an operating temperature of 1500°F the unit is ready for the process airstream. As the process airstream enters the ceramic heat transfer beds, the heated ceramic media preheats the process airstream to its oxidation temperature. Oxidation of the airstream occurs when the autoignition of the hydrocarbon is reached. At this point the heat released by the oxidation of the process hydrocarbons is partially absorbed by the inlet ceramic heat transfer bed. The heated air passes through the retention chamber, and the heat is absorbed by the outlet ceramic heat transfer bed.

During the normal mode of operation of the system the process air enters the RTO system fan and passes through the inlet diverter valve, where the process air is forced into the bottom of the left ceramic heat transfer bed. As the process air rises through the ceramic heat transfer bed, the temperature of the process stream will rise. The tops of the beds are controlled to a temperature of 1500°F. The bottoms of the beds will vary depending upon the temperature of the air that is coming in. If it is assumed that the process air is at ambient conditions or 70°F, then as the air enters the bottom of the bed, the bottom of the bed will approach the inlet air temperature of 70°F. The entering air is heated and the media is cooled. As the air exits the ceramic media it will approach 1500°F. The process air then enters the second bed at 1500°F, and now the ceramic media recovers the heat from the air and increases in temperature. At a fixed time interval (usually 4 to 5 minutes), or based on thermocouple control, the diverter valves switch and the process air is directed to enter the bed on the right and exit the bed on the left. Prior to valve switching the air heated the right bed, and now this bed is being cooled. The cooling starts at the bottom and continues upward because the media is hot and the energy is transferred. The process air then goes through the purification chamber and exits through the second bed. When the valves are switched, whatever organics had not been destroyed prior to the flow being reversed are then exhausted out of the stack. In addition, the rapidity of switching or closure of the valves is critical to minimize the bypass of unoxidized organics. If the emissions versus time were plotted, the graph would reflect a very low exhaust concentration level, but whenever the diverter valve switches an organic pulse occurs in the exhaust stream. Since the valves shift every 4 minutes these pulses reduce the overall destruction efficiency of the organics. Several methods of processing the pulse achieve higher destruction efficiencies.

The retention chamber and heat transfer chamber are fabricated of reinforced carbon steel exterior and ceramic fiber lining. The thickness of the ceramic fiber lining is based on the required destruction temperature of the organics and the desired outside shell temperature. The ceramic heat exchange media can be of various types including ceramic saddles, tye pacs, or structured packing. The ceramic structured packing is a recent development in the industry, reflecting lower pressure drops for equivalent heat transfer. A reinforced carbon steel structure is provided to support the loads of the oxidizer chambers and the structured packing support grid, and the wind or earthquake loads. The packing support grid is fabricated of stainless steel and is designed to support the structured packing. If organic particulates in the process exhaust build up on the "cold" surfaces at the bottom of the oxidizer, the process must be shut down, and a volatilization of these organics or a "bake-out" is required. When a bake-out is activated, the flow diverter valves will stay in one position until the exhaust air temperature from the outlet bed reaches 850°F. At this temperature, most organic oils will volatilize, as in a self-cleaning oven. When the first outlet bed reaches 850°F, the flow diverter valves will switch and stay in position until the outlet temperature of the second bed reaches 850°F.

Regenerative Catalytic Oxidizer (RCO). A catalytic oxidizer design uses Gas Hourly Space Velocity (GHSV), which relates the amount of air to be processed and the volume of catalyst required. The GHSV is the inverse of the residence time. We note also that lower temperature translates into lower fuel costs, assuming the same heat exchanger efficiency. It takes energy to take the airstream from whatever temperature it enters up to its operating temperature. Even though a heat exchanger will help reduce net fuel costs, the lower the operating temperature, the lower the fuel costs.

Rotor Concentrator/Thermal Oxidizer. For high-volume airstreams with relatively low organic concentration, the energy required to bring the process stream to thermal oxidation temperature can be significant. A technology using a rotor concentrator in a two-step process has proven to be effective in high-volume, low-concentration airstreams.

The first part of that system consists of a slowly rotating concentrator wheel, utilizing zeolites or carbon deposited on a substrate, which adsorbs the organics as they are exhausted from the original process and passed through the wheel. A sector of the concentrator wheel is partitioned off from the main section of the rotor and clean heated air is passed through this section to desorb the organics. The air volume of clean heated air varies for applications, but is roughly 10% of the original airstream volume.

As the clean heated air passes through the rotor section, the organics are desorbed, resulting in a process stream that is 10% in volume of the original process stream and 10 times the original concentration of organics. The desorption airstream will drive off all the organics in that particular sector of the wheel, and as the shell continues to turn the organics are desorbed from the entire wheel. As the low-volume airstream exits, it is processed through a thermal recuperative oxidizer in the second part of the system. Consequently, what is achieved is that a large airstream with low concentration, which requires considerable energy to heat up, has been reduced to as low-volume airstream with a high concentration, which is significantly less costly to process.

A secondary fan draws the air and forces it through the secondary heat exchanger, where the reduced air volume temperature is raised to the required desorption temperature. The preheated air is then used to desorb the air in another portion of the wheel. As the air exits the desorption section the organic concentration is approximately 10 times the concentration of the original process stream. This low-volume, higher concentration stream then enters the induced draft section of a catalytic or thermal recuperative oxidizer, where the organics are destroyed.

The total hydrocarbon reduction efficiency for the rotor/concentrator is the adsorption efficiency of the rotor/concentrator times the destruction efficiency of the oxidizer.

Flare and Burners. Certainly the oldest and still widely used technology through some parts of the world is flaring. Flares are used in the petroleum, petrochemical, and other industries that require the disposal of waste gases of high concentration of both a continuous or intermittent basis. As other thermal oxidation technologies, the three Ts of combustion of time, temperature, and turbulence are necessary to achieve adequate emission control.

Flares ideally burn waste gas completely and smokelessly. Two types of flares are normally employed. The first is called the open flare, the second is called the enclosed flare. The major components of a flare consist of the burner, stack, water seal, controls, pilot burner, and ignition system. Flares required to process variable air volumes and concentrations are equipped with automatic pilot ignition systems, temperature sensors, and air and combustion controls.

Open flares have a flare tip with no restriction to flow, the flare tip being the same diameter of the stack. Open flares are effectively a burner in a tube. Combustion and mixing of air and gas take place above the flare with the flame being fully combusted outside of the stack.

Enclosed flares are composed of multiple gas burner heads placed at ground level in a stacklike enclosure that is usually refractory or ceramic lined. Many flares are equipped with automatic damper controls that regulate the supply of combustion air, depending on temperature, which is monitored upstream of the mixing, but inside the stack. This class of flare is becoming the standard in the industry due to its ability to more effectively control emissions. Requirements on emissions includes carbon monoxide limits and minimal residence time and temperature. Exhaust gas temperatures may vary from 1000 to 2000°F.

Ceramic Filter. The ceramic filter collects, volatilizes, and destroys particulates and condensible organics emitted from industrial process streams, such as paint spray, lost foam casting, condensible organics, tenter frames, and cured rubber operations. It may be used independently as a hot or cold filter media or coupled with an oxidization module for total odor and VOC control. The ceramic filter is designed based on exhaust airflow volume, type of contaminant, and desired collection efficiency. During operation, the process stream, containing hot or cold particulate-laden air, condensible organics, or VOCs, is drawn into the ceramic filter. The process stream passes over a ceramic matrix selected on particle size and collection efficiency considerations. The ceramic matrix is periodically heated by a natural gas burner, whereby any organics collected on the ceramic matrix are volatilized. Any non-organics collected on the ceramic matrix are converted to inorganic ash and retained in the unit dropout chamber. If required, the volatilized organics can then be processed through a catalyst bed. As in the catalytic oxidizer, these volatiles would be converted to CO_2 and H_2O.

RECOMMENDED RESOURCES

1. AAF International, Inc., www.aafint/com/equipmentl, Core Products Information, last updated December 1992, Air & Waste Management Association (AWMA).

2. AWMA, *Air Pollution Engineering Manual*. Van Nostrand Reinhold, New York, 1999.

3. Andriola, 1999. T. Andriola, Fisher-Klosterman, Inc., personal communication with Eric Albright, October 14, 1999.

4. Avallone, 1996. *Marks' Standard Handbook for Mechanical Engineers*, edited by Eugene Avallone and Theodore Baumeister, McGraw-Hill, New York, 1996.

5. Avallone, 1996. *Marks' Standard Handbook for Mechanical Engineers,* edited by Eugene Avallone and Theodore Baumeister, McGraw-Hill, New York, 1996.

6. AWMA, 1992. Air & Waste Management Association, *Air Pollution Engineering Manual*, Van Nostrand Reinhold, New York, 1992.

7. AWMA, 1992. Air & Waste Management Association, *Air Pollution Engineering Manual*, Van Nostrand Reinhold, New York, 1992.

8. AWMA, 1992. Air & Waste Management Association, *Air Pollution Engineering Manual,* Van Nostrand Reinhold, New York, 1992.

9. AWMA, 1992. Air & Waste Management Association, *Air Pollution Engineering Manual,* Van Nostrand Reinhold, New York, 1992.

10. Cheremisinoff and Young, 1977. Paul N. Cheremisinoff and Richard A. Young, *Air Pollution Control and Design Handbook, Vol. 2,* Marcel Dekker Publishers, New York, 1977.

10. *Control Cost Manual, Fifth Edition,* EPA 453/B-96-001, Research Triangle Park, NC, February 1996.

12. Cooper and Alley, 1994. C. D. Cooper and F. C. Alley, *Air Pollution Control: A Design Approach, Second Edition,* Waveland Press, Prospect Heights, IL, 1994.

13. Cooper and Alley, 1994. C. D. Cooper and F. C. Alley, *Air Pollution Control: A Design Approach, Second Edition,* Waveland Press, Prospect Heights, IL, 1994.

14. Cooper and Alley, 1994. C. D. Cooper and F. C. Alley, *Air Pollution Control: A Design Approach, Second Edition,* Waveland Press, Prospect Heights, IL, 1994.

15. Corbitt, 1990. *Standard Handbook of Environmental Engineering,* edited by Robert Corbitt, McGraw-Hill, New York, 1990.

16. Davis, 1999. W. Davis, Professor and Coordinator, Environmental Engineering Program, Department of Civil and Environmental Engineering, University of Tennessee, personal communication with Eric Albright, October 28, 1999.

17. EPA, 1998a. U.S. EPA, Office of Air Quality Planning and Standards, *Stationary Source Control Techniques Document for Fine Particulate Matter,* EPA-452IR-97-001, Research Triangle Park, NC, October 1998.

18. EPA, 1982. U.S. EPA, Office of Air Quality Planning and Standards, *Control Techniques for Particulate Emissions from Stationary Sources, Volume 1,* EPA-450/3-81-005a, Research Triangle Park, NC, September 1982.

19. EPA, 1996. U.S. EPA, Office of Air Quality Planning and Standards, *OAQPS Control Cost Manual, Fifth Edition,* EPA 453/B-96-001, Research Triangle Park, NC, February, 1996.

20. EPA, 1996. U.S. EPA, Office of Air Quality Planning and Standards, *OAQPS Control Cost Manual, Fifth Edition,* EPA-453/B-96-001, Research Triangle Park, NC, February 1996.

21. EPA, 1996. U.S. EPA, Office of Air Quality Planning and Standards, *OAQPS Control Cost Manual, Fifth Edition,* EPA-453/B-96-001, Research Triangle Park, NC, February 1996.

22. EPA, 1996. U.S. EPA, Office of Air Quality Planning and Standards, *OAQPS Control Cost Manual," Fifth Edition,* EPA-453/B-96-001, Research Triangle Park, NC, February 1996.

23. EPA, 1997. U.S. EPA, Office of Air Quality Planning and Standards, *Compilation of Air Pollutant Emission Factors. Volume I, Fifth Edition,* Research Triangle Park, NC, October 1997.

24. EPA, 1997. U.S. EPA, Office of Air Quality Planning and Standards, *Compilation ofAir Pollutant Emission Factors, Volume I, Fifth Edition,* Research Triangle Park, NC, October 1997.

25. EPA, 1997. U.S. EPA, Office of Air Quality Planning and Standards, *Compilation of Air Pollutant Emission Factors, Volume I, Fifth Edition,* Research Triangle Park, NC, October 1997.

26. EPA, 1997. U.S. EPA, Office of Air Quality Planning and Standards, *Compilation of Air Pollutant Emission Factors, Volume I, Fifth Edition,* Research Triangle Park, NC, October 1997.

27. EPA, 1998. U.S. EPA, Office of Air Quality Planning and Standards, *Stationary Source Control Techniques Document for Fine Particulate Matter,* EPA-452/R-97-001, Research Triangle Park, NC, October 1998.

28. EPA, 1998. U.S. EPA, Office of Air Quality Planning and Standards, *Stationary Source Control Techniques Document for Fine Particulate Matter,* EPA-452/R-97-001, Research Triangle Park, NC, October 1998.

29. EPA, 1998. U.S. EPA, Office of Air Quality Planning and Standards, *Stationary Source Control Techniques Document for Fine Particulate Matter,* EPA-452/R-97-001, Research Triangle Park, NC, October 1998.

30. EPA, 1998. U.S. EPA, Office of Air Quality Planning and Standards, *Stationary Source Control Techniques Document for Fine Particulate Matter,* EPA-452/R-97-001, Research Triangle Park, NC, October 1998.

31. EPA, 1998a. U.S. EPA, Office of Air Quality Planning and Standards, *OAQPS Control Cost Manual,"* Fifth Edition, Chapter 5, EPA-453/B-96-001, Research Triangle Park, NC, December 1998.

32. EPA, 1998b. U.S. EPA, Office of Air Quality Planning and Standards, *Stationary Source Control Techniques Document for Fine Particulate Matter* EPA-452/R-97-001, Research Triangle Park, NC, October 1998.

33. EPA. 1998b. U.S. EPA, Office of Air Quality Planning and Standards, *OAQPS Control Cost Manual, Fifth Edition,* Chapter 5, EPA-453/B-96-001, Research Triangle Park, NC, December 1998.

34. Flynn, 1999. Brian Flynn, Beltran Associates, Inc., personal communication with Eric Albright, February 5, 1999.

35. Heumann, 1997. W. L. Heumann, *Industrial Air Pollution Control Systems*, McGraw-Hill Publishers, Inc., Washington, DC, 1997.

36. ICAC, 1999. Institute of Clean Air Companies (ICAC), www.icac.com, Control Technology Information: Electrostatic Precipitator, page last updated January 11, 1999.

37. ICAC, 1999. Institute of Clean Air Companies, www.icac.com, Control Technology Information; Fabric Filters, page last updated January 11, 1999.

38. ICAC, 1999. Institute of Clean Air Companies, www.icac.com, Control Technology Information: Fabric Filters, page last updated January 11, 1999.

39. IFF, 1999. Industrial Filter Fabric, Inc., www.filters.com, Product Bulletin 003; Cartridge Filters, December 1999.

40. Josephs, 1999. D. Josephs, Equipment Product Manager, AAF International, personal communication with Eric Albright, October 28, 1999.

41. Mycock, 1995. J. Mycock, J. McKenna, and L. Theodore, *Handbook of Air Pollution Control Engineering and Technology*, CRC Press, Boca Raton, FL, 1995.

42. Parsons, 1999. B. Parsons, Sterling Systems, Inc., personal communication with Eric Albright, October 26, 1999.

43. Perry, 1984, *Perry's Chemical Engineer's Handbook, Sixth Edition,* edited by Robert Perry and Don Green, McGraw-Hill, New York, 1984.

44. Perry, 1984. "Perry's Chemical Engineers' Handbook, *Sixth Edition*, edited by Robert Perry and Don Green, McGraw-Hill, New York, 1984.

45. STAPPA/ALAPCO, 1996. State and Territorial Air Pollution Program Administrators and the Association of Local Air Pollution Control Officials, *Controlling Particulate Matter Under the Clean Air Act: A Menu of Options*, STAPPA/ALAPCO, Washington, D.C., July 1996.

46. STAPPA/ALAPCO, 1996. State and Territorial Air Pollution Program Administrators and Association of Local Air Pollution Control Officials. *Controlling Particulate Matter Under the Clean Air Act: A Menu of Options*, July 1996.

47. STAPPA/ALAPCO, 1996. State and Territorial Air Pollution Program Administrators and Association of Local Air Pollution Control Officials, *Controlling Particulate Matter Under the Clean Air Act: A Menu of Options*, July 1996.

48. STAPPAIALAPCO, 1996. State and Territorial Air Pollution Program Administrators and Association of Local Air Pollution Control Officials, *Controlling Particulate Matter Under the Clean Air Act: A Menu of Options*, July 1996.

49. Torit, 1999. Torit Products, a subsidiary of Donaldson Company, www.torit.com/Products, Industrial Dust Collection Systems, last updated December 1999.

50. Vatavuk, 1990. W.M. Vatavuk, *Estimating Costs of Air Pollution Control*, Lewis Publishers, Chelsea, MI, 1990.

51. Wark, 1981. Kenneth Wark and Cecil Warner, *Air Pollution: Its Origin and Control*, Harper Collins, New York, 1981.

6

Integration of Gasification Technologies

INTRODUCTION

Gasification technologies offer the potential to provide a stable energy supply. Gasification-based systems offer the promise of high efficiency with a green technology basis. Additionally they offer flexibility in the production of a broad spectrum of products, including electricity, fuels, chemicals, hydrogen, and steam.

This chapter provides an overview of the technical status of integrated gasification technologies as applied to integrated gasification combined cycle (IGCC) for electricity. Gas turbine (GT), fuels cells, and overall system integration requirements within gasification processes are discussed within the context of fully integrated systems.

To understand market opportunities as they are evolving, the technology status of major components within current and future IGCC systems must be understood.

ROLE OF COAL GASIFICATION

Coal accounts for 90% of the proven energy reserves in the United States, and is a source of fuel for nearly half of the nation's electricity, typically via combustion in pulverized coal (PC)-fired boilers. While the U.S. is rich in coal reserves, environmental and siting issues pose major obstacles to tapping its full potential for energy security. Nearly 63% of sulfur emissions and 19% nitrogen emissions in the U.S. are derived from PC-fired power plants. Coal-based power plants have inherent fugitive particulate emissions that pose long-term health risks, as well as costly solids handling issues that have given rise to numerous nuisance complaints and objections by communities. Further, the large capital investment requirements for coal-fired plants, which operate at relatively low thermodynamic efficiencies, is a serious concern in light of more modern global climate change policy alternatives.

In contrast, coal gasification produces a relatively clean fuel. Further environmental credits can be gained by augmenting coal with opportunity fuels that contain fewer toxic components, such as bio-materials, municipal solid wastes, and some petroleum refinery wastes. These feedstocks not only reduce overall societal burdens, but they can, in some instances, improve overall process economics. Gasification technologies further lend themselves to process features that are more effective at cleaning the product gas at the "fuel gas" stage, thereby removing many of the air pollutants potentially under optimum economic conditions; for example, sulfur impurities resulting in SO_X can be typically removed to efficiencies in excess of 95%. The cleaned product gas can then be burned using low nitrous (NO_X) combustion technologies to produce power in electrical generating systems.

Coal gasification represents the next generation of coal-based energy production. Instead of combusting coal directly, gasification reacts coal with steam and carefully controlled amounts of air or oxygen under high temperatures and pressures. The heat and pressure dissociate the chemical bonds in coal, generating chemical reactions with the steam and oxygen to form a gaseous mixture, which includes large amounts of hydrogen and carbon monoxide.

As already noted, hydrogen is of particular interest from both an environmental and thermodynamic standpoint because it is the cleanest and highest efficiency burning fuel. But also, pollutant-forming impurities and greenhouse gases can be separated from the gaseous stream, referred to as producer gas or syngas. As much as 99% of sulfur and other pollutants can be removed and processed into commercial products (e.g., feedstock chemicals and fertilizers). Unreacted solids can be collected and marketed as a co-product (e.g., slag, which can be used in road building). The primary product, fuel-grade coal-derived gas, rivals clean natural gas in environmental quality. The basic gasification process can also be applied to virtually any carbon-based feedstock. Common feedstock fuels are biomass (e.g., wood chips, agricultural waste, switchgrass), petroleum coke, municipal waste, farm animal manure, wastewater treatment sludges, or blends of these fuels.

Coal gasification offers a more efficient approach to electricity generation than conventional coal-burning power plants. In a conventional plant, heat from the coal furnace is used to boil water, creating steam for a steam-turbine generator. By contrast, a gasification-based power plant uses the hot, high-pressure coal gases exiting a gasifier to power a gas turbine, in the same manner as natural gas. Hot exhaust from the gas turbine may then be fed into a conventional steam turbine, producing a second source of power. This dual, or "combined-cycle," configuration of turbines is not possible with conventional coal combustion and, therefore, offers significant improvements in power plant efficiencies. Conventional combustion plants are typically 33 to 35% efficient (fuel-to-electricity). But coal gasification offers the prospects of boosting efficiencies to 45 to 50% in the short term and potentially to nearly 60% with technological advancements. Higher efficiencies translate into better economics and inherent reductions in greenhouse gases.

Like natural gas, the producer gas from coal is a clean fuel. Additionally, it is a rich source of chemicals. Coal-derived gas can also be recombined into liquid fuels, including high-grade transportation fuels, and a range of petrochemicals that serve as feedstock workhorses in the chemicals and refining industries. In contrast to conventional combustion, carbon dioxide exits a coal gasifier in a concentrated stream rather than diluted in a high volume of flue gas. This allows the carbon dioxide to be captured more effectively and then used for commercial purposes or sequestered. From a historical standpoint, industry's interest in gasification has been to produce fuels, chemicals, and fertilizers, with technology origins dating back to WWII. And although gasification is used in refineries and chemical plants throughout the world, the technology is still in the demonstration phase for electric power generation.

The generalized concept of gasification as applied to electricity generation is the integrated gasification combined cycle. In the IGCC system, clean fuel gas is burned in a gas turbine engine, which generates electricity. The exhaust gases from the GT remain hot enough to produce steam to effectively drive a conventional steam cycle, thus producing more electricity. The IGCC system is referred to as being "integrated" because working fluids (air or steam) flow between the gasification, GT, and conventional steam cycle subsystems. This enables exchange and utilization of energy in a highly opportunistic manner, resulting in higher overall system performance compared to conventional technology. Coal-to-electricity conversion efficiencies for IGCC systems normally exceed 40%, with 50 to 55% levels projected for configurations that are coupled with advanced gas turbines. In contrast, PC-fired generating plants are typically in the low to mid-30% league. In addition, an IGCC electric generating plant is expected to have at least 25 to 35% fewer air emissions, ash to be disposed of, and coal to be purchased compared to state-of-the-art PC-fired plants.

The U.S. Department of Energy (USDOE) has assisted in technology development and demonstrations for both large- and small-scale programs. Semi-commercial experience and the most reliable data on likely commercial investment costs appear to be for large-scale operations (above 200 MWe). Therefore, in assessing the small-scale applications range, reference to large-scale experience is appropriate.

Under the original Clean Coal Technology Program, utilities built and successfully operated coal gasification power plants. One of these is the Tampa Electric Integrated Gasification Combined-Cycle Project, which is a large-scale program.

This is an advanced electric power generation IGCC demonstration in a greenfield commercial electric utility application at the 250 MWe size using an entrained-flow, oxygen-blown gasifier with full heat recovery, conventional cold-gas cleanup, and an advanced gas turbine with nitrogen injection for power augmentation and NO_x control. In this process coal/water slurry and oxygen are reacted at high temperature and pressure to produce a medium-Btu syngas in a Texaco gasifier. Molten ash flows out of the bottom of the gasifier into a water-filled sump, where it is forms a solid slag. The syngas moves from the gasifier to a high-temperature heat-recovery unit, which cools the syngas while generating high-pressure steam. The cooled gases flow to a water wash for particulate removal. Next, a COS hydrolysis reactor converts one of the sulfur species in the gas to a form that is more easily removed. The syngas is then further cooled before entering a conventional amine sulfur-removal system. The amine system keeps SO_2 emissions below 0.15 lb/106 Btu (97% capture). The cleaned gases are then reheated and routed to a combined-cycle system for power generation. The following Web site provides more in-depth discussions plus a process flowsheet: http://www.teco.net/teco/TEPlkPwrStn.html.

Some basic facts about the Tampa Electric IGCC Project are:

- A GE MS 7001FA gas turbine generates 192 MWe.
- The plant heat rate is 9350 Btu/kWh (HHV).
- SO_2 reduction of 95% achieved.
- The gasifier operated more than 29,000 hours and processed coal at a rate of 2,300 tons/day.
- The combustion turbine operated over 28,000 hours to produce over 8.6 million MWh of electricity on syngas.
- Carbon burnout exceeds 95%.
- Total cost of the Tampa Electric IGCC Project is $303 million, or $1213/kW.
- The project successfully demonstrated the commercial application of Texaco coal gasification in conjunction with electric power generation.

- DOE estimates that future IGCC power plants, based on mature and improved technology, will cost in the range of $900 to 1250/kW (1999$) depending on the degree to which existing equipment and infrastructure can be utilized.
- Heat rate ultimately is expected to be in the range of 7000–7500 Btu/kWh (46–49%; HHV).

The Wabash River Coal Gasification Repowering Project is also an IGCC demonstration. The project objective is to demonstrate utility repowering with a two-stage, pressurized, oxygen-blown, entrained-flow IGCC system, including advancements in the technology relevant to the use of high-sulfur bituminous coal; and to assess long-term reliability, availability, and maintainability (RAM) of the system at a fully commercial scale. The Destec, now E-Gas Technology™, process features an oxygen-blown, continuous-slagging, two-stage, entrained-flow gasifier. Coal is slurried, combined with 95% pure oxygen, and injected into the first stage of the gasifier, which operates at 2600°F/400 psig. In the first stage, the coal slurry undergoes a partial oxidation reaction at temperatures high enough to bring the coal's ash above its melting point. The fluid ash falls through a tap hole at the bottom of the first stage into a water quench, forming an inert vitreous slag.

The syngas flows to the second stage, where additional coal slurry is injected. This coal is pyrolyzed in an endothermic reaction with the hot syngas, to enhance syngas heating value and improve efficiency. The syngas then flows to the syngas cooler, essentially a firetube steam generator, to produce high-pressure saturated steam. After cooling in the syngas cooler, particulates are removed in a hot/dry filter and recycled to the gasifier. The syngas is further cooled in a series of heat exchangers. The syngas is water-scrubbed to remove chlorides and passed through a catalyst that hydrolyzes carbonyl sulfide into hydrogen sulfide. Hydrogen sulfide is removed in the acid gas removal system using MDEA-based absorber/stripper columns. A Claus unit is used to produce elemental sulfur as a salable by-product. The "sweet" gas is then moisturized, preheated, and piped to the power block. The power block consists of a single 192 MWe General Electric MS7001FA (Frame 7FA) gas turbine, a Foster Wheeler single-drum heat-recovery steam generator with reheat, and a 1952 vintage Westinghouse reheat steam turbine. The following Web site provides further information plus a process flowsheet: http://www.lanl.gov/projects/cctc/factsheets/wabsh/wabashrdemo.html.

Some basic facts about the Wabash River Coal Gasification Repowering Project are:

- SO_2 capture efficiency was greater than 99%.
- NOx emissions were 0.15 lb/106 Btu, which meets the 2003 target emission limits for ozone non-attainment areas, or 1.09 lb/MWh, which exceeds the New Source Performance Standard of 1.6 lb/MWh.
- Particulate emissions were below detectable limits.
- Carbon monoxide emissions, averaging 0.05 lb/106 Btu, were well within industry standards.
- Coal ash was converted to a low-carbon vitreous slag, impervious to leaching and valued as an aggregate in construction or as grit for abrasives and roofing materials; trace metals from petroleum coke were also encased in an inert vitreous slag.
- The IGCC unit operated on coal for over 15,000 hrs, processed over 1.5 million tons of coal, and produced over 23 trillion Btu of syngas and 4 million MWh of electricity.
- The overall cost of the IGCC plant was $417 million, which equates to about $1,590/kW in 1994 dollars. For an equivalent greenfield project the cost was estimated at $1,700/kW.
- Capital cost estimates for a new 285 MWe (net) greenfield IGCC plant incorporating lessons learned, technology improvements, and a heat rate of 8526 Btu/kWh are $1,318/kW (2000$) for a coal-fueled unit and $1,260 (2000$) for a petroleum coke-fueled unit.

There are three main types of gasification technology: entrained flow, fluidized bed, and moving bed. All of these can be used to gasify coal, and gasifier selection will depend on coal characteristics and plant size. In an entrained-flow gasifier pulverized coal flows co-currently with the oxidant (usually oxygen). These gasifiers are characterized by very high (usually > 1000°C) and uniform temperatures and short residence times. Ash melts and is removed as a liquid slag. Entrained flow gasifiers have been selected for most of the coal-fired systems currently in operation or under construction. Examples of entrained-flow gasifiers include Texaco, Shell, and Prenflo.

In a fluidized-bed gasifier the fuel is suspended in the upward flowing gas stream. Operating temperatures are maintained well below 1000°C to prevent ash becoming sticky and sintering the bed. Fluidized bed gasifiers can accept a wide range of fuel types including high-ash, high-ash fusion temperature coals such as those prevalent in the Southern Hemisphere. Examples of fluidized-bed gasifiers include the High Temperature Winkler and Mitsui Babcock Energy Limited's spouted gasifier (licensed from British Coal).

The USDOE experience has proven that there are three types of syngas, each with different calorific values. These are summarized in Table 6.1.

Chapter 1 provided an overview of the most recent developments in gasification technologies. The three primary types of generic gasifiers are entrained flow, fluidized bed, and moving bed.

Entrained-flow gasifiers are most developed at industrial and demonstration scales with, among others, IGCC plants in operation in the United States (Tampa Electric with the Texaco technology, Wabash River with the Global Energy technology) and two IGCC plants in Europe (Buggenum with the Shell technology and Puertollano with the Prenflo technology). New gasification plants based on other entrained-flow gasifier technologies are about to be built in Japan (an IGCC plant with the MHI technology and the Hitachi technology for the Eagle project). The entrained-flow gasifiers are all oxygen-blown slagging gasifiers, producing medium calorific value syngas. As such, they can be used either for power

TABLE **6.1**   Types of Syngas From Coal Gasification Technologies

Heating Value	Calorific Value (MJ/m³)	Suitability	Other Applications or Restrictions
Low	3.8–7.6	Fuel gas for gas turbines as in IGCC	Can be used for smelting and reduction of iron ore. Due to high nitrogen content, cannot be used as a substitute for natural gas replacement or for chemical synthesis.
Medium	10.5–16	Fuel gas for gas turbines (IGCC) and as a replacement for natural gas or chemical substitute. Can be used as a feedstock to Fisher-Tropsch synthesis, methanation, methanol, and ammonia production. H₂ production for fuel cell (high temperature solid fuel cell and molten carbonate) is also possible.	
High	>21	Substitute for natural gas	

generation or chemical synthesis. They are regarded as the most versatile type of gasifiers, as they can accept both solid and liquid fuels and operate at high temperatures (above ash slagging temperatures) to ensure high carbon conversion and produce a syngas free of tars and phenols. However, high temperatures have an impact on burners and refractory life and require the use of expensive materials of construction as well as the use of sophisticated high-temperature heat exchangers to cool the syngas below the ash softening temperature. Entrained flow gasifiers are usually not recommended for high-ash-content coals in order to avoid fouling and minimize corrosion problems and, as it is a slagging gasifier, for coals with a high ash fusion temperature.

Fluidized-bed gasifiers are used in six different types of gasification processes. However, few of those processes have been developed at demonstration scale. There is only one IGCC demonstration project using the KRW technology (Pinon Pine) in the United States, and that demonstration plant has faced numerous problems, apparently not related to the quality of the coal used, since its commissioning. There is presently a project in the Czech Republic to develop an IGCC based on the HTW gasification technology to replace old moving-bed gasifiers.

Fluidized-bed gasifiers can only operate with solid crushed fuels. They can differ in ash discharge conditions and can be ash-dry or ash-agglomerated. The agglomerated ash operation improves the ability to gasify high-rank coals more efficiently. Conventional dry ash operation has traditionally operated with low-rank coals. One of the main advantages of this type of gasifier is that they can operate at variable loads, giving them a high turndown flexibility. However, they operate at lower temperatures than the other two types of generic gasifiers, with the consequence that carbon can be found in fly ash recovered in the particulate removal units (cyclones). This leads to lower cold gas efficiencies in fluidized-bed gasifiers compared to moving-bed gasifiers. Hybrid systems, such as the ABGC, which includes the gasification of coal followed by the combustion of the char, can solve this problem and increase carbon conversion, leading to higher cold gas efficiency. Fluidized beds can also process coals with a relatively high sulfur content that can be partly (up to 90%) retained in the bed by sorbents such as limestone and hence reduce the presence of corrosive H_2S in the downstream equipment. As fluidized-bed gasifiers operate at a lower temperature than entrained-flow gasifiers, they have the advantage of having a cheaper cooling system.

There are only three types of processes using moving-bed gasifiers (BGL, BHEL, Lurgi) developed at industrial scale. They are the most mature technologies of the three types of generic gasifiers. There is presently one coal IGCC plant operating with moving-bed gasifiers, which are going to be replaced soon by fluidized-bed gasifiers, and another one located in Germany (Schwarpe Pumpe, BGL technology) for the processing of wastes and coal. Chemical gasification plants based on moving-bed technologies are at present operating all over the world, with the biggest plants being located in South Africa. Finally there is a project sponsored by the USDOE in the United States (Kentucky) to build a gasification plant, which includes a molten carbonate fuel for power generation. Moving-bed gasifiers can be either slagging (BGL) or dry ash (Lurgi, BHEL). They are only suitable for solid fuels, and their main requirements are efficient heat and mass transfer between solids and gases within the bed. That requires good bed permeability and consequently the control of coal particle size. As a consequence they cannot process coal fines and are also not recommended for caking coals. The main advantage of moving-bed gasifiers is that they do not require the use of expensive syngas coolers, as typical gas exit temperatures are of the order of 400 to 500°C due to countercurrent flows. Nevertheless, the temperature at the top of the gasifier is usually not high enough to break down the tar, phenols, oils, and low-boiling-point hydrocarbons produced in the pyrolysis zone and carried out with the gasifier product gas. Recent design changes incorporate recycling, which helps to consume these by-products to extinction.

Gasification technologies can be adapted to process all types of coals for the production of electricity (IGCC), the production of hydrogen for use in fuel cells (FCs), and also for the cogeneration of chemicals. Entrained-flow gasifier processes are very flexible and can process all coal ranks. However, they have a preference for coals with a low ash content (10% or less), with some of the processes requiring a minimum ash content. They are all slagging gasifiers and are usually designed to process coals with ash fusion temperatures lower than 1400°C and slag viscosities lower than 15 Pas at that temperature. Gasification processes using fluidized-bed gasifiers are usually recommended for low-rank coals with high reactivities and particularly for dry ash gasifiers, coals with high ash fusion temperatures. They can also process coals with a relatively high sulfur content that can be partly (up to 90%) retained in the bed by sorbents. Gasification processes based on moving-bed gasifiers have good flexibility in terms of coal rank but they cannot process coal fines and strongly caking coals. The slagging version of moving-bed gasifiers is also not recommended, as entrained-flow gasifiers for coals have very high ash content and very high ash melting points. Tapping temperatures recommended should be lower than 1400°C for slag viscosities lower than 5 Pas.

Few coal gasification plants have yet reached commercial development, and they still have to compete with cheaper technologies such as conventional pulverized coal-fired plants and natural gas for the production of electricity.

GAS TURBINE TECHNOLOGIES

Gas turbines are one of the cleanest means of generating electricity, with emissions of NO_X from some large turbines in the single-digit ppm range, either with catalytic exhaust cleanup or lean pre-mixed combustion. Because of their relatively high efficiency and reliance on natural gas as the primary fuel, gas turbines emit substantially less CO_2 per kWh generated than any other technology in general commercial use.

A gas turbine produces a high temperature, high-pressure gas working fluid, through combustion, to induce shaft rotation by impingement of the gas upon a series of specially designed blades. The shaft rotation drives an electric generator and a compressor for the air used by the gas turbine. Many turbines also use a heat exchanger (called a "recuperator") to impart turbine exhaust heat into the combustor's air/fuel mixture. Gas turbines can be used in three primary operating configurations:

1. Simple cycle operation, which is a single gas turbine producing power only
2. Combined heat and power (CHP) operation, which is a simple cycle gas turbine with a heat recovery heat exchanger that recovers the heat in the turbine

exhaust and converts it to useful thermal energy, usually in the form of steam or hot water

3. Combined-cycle operation in which high-pressure steam is generated from recovered exhaust heat and used to create additional power using a steam turbine

In addition, some combined cycles extract steam at an intermediate pressure for use in industrial processes and are combined-cycle CHP systems.

Machines are available in sizes ranging from 500 kW to 250 MW. The most efficient commercial technology for central station power-only generation is the gas turbine-steam turbine combined-cycle plant, with efficiencies approaching 60% (LHV). Simple-cycle gas turbines for power-only generation are available with efficiencies approaching 40% (LHV). Gas turbines have traditionally been used by utilities for peaking capacity. However, with changes in the power industry and advancements in the technology, the gas turbine is now being increasingly used for baseload power.

Efficiency and emissions are critical to the CHP market. The high-temperature exhaust from gas turbines provides great flexibility in meeting thermal needs at the site. The most common use of this energy is for steam generation in unfired or supplementary fired heat-recovery steam generators (HRSGs). However, the gas turbine exhaust gases can also be used as a source of direct process energy, for unfired or fired process fluid heaters, or as preheated combustion air for power boilers. Since gas turbine exhaust is oxygen-rich and can support additional combustion, even greater flexibility can be achieved through supplementary firing of the HRSG. A duct burner can be fitted within the HRSG to further raise the exhaust temperature and increase the amount and temperature of steam generated at incremental efficiencies of 90% or greater. Gas turbines are ideally suited for CHP applications because their high-temperature exhaust can be used to generate process steam at conditions as high as 1250 psig and 900°F.

Gas turbine systems operate on the thermodynamic cycle known as the Brayton cycle. In a Brayton cycle, atmospheric air is compressed, heated, and then expanded, with the excess of power produced by the expander over that consumed by the compressor used for power generation. The power produced by an expansion turbine and consumed by a compressor is proportional to the absolute temperature of the gas passing through the machine. It is advantageous to operate the expansion turbine at the highest practical temperature consistent with economic materials and internal blade cooling technology, and to operate the compressor with inlet air flow at as low a temperature as possible. As technology advances permit higher turbine inlet temperature, the optimum pressure ratio also increases. Higher temperature and pressure ratios result in higher efficiency and specific power. Thus the general trend in gas turbine advancement has been towards a combination of higher temperatures and pressures. While such advancements increase the manufacturing cost of the machine, the higher value, in terms of greater power output and higher efficiency, provides net economic benefits. The industrial gas turbine is a balance between performance and cost that results in the most economic machine for both the user and manufacturer. The temperature versus cost tradeoff also involves emissions, which generally increase with increased firing temperature. Often emissions regulations result in a practical upper limit on temperature before materials considerations do. The primary components of a simple cycle gas turbine are shown in Figure 6.1.

There are two general types of machines based on current technology. Aeroderivative gas turbines for stationary power are adapted from their jet and turboshaft aircraft engine counterparts. While these turbines are lightweight and thermally efficient, they are usually more expensive than products designed and built exclusively for stationary applications. The largest aeroderivative generation turbines available are 40 to 50 MW in capacity. Many aeroderivative gas turbines for stationary use operate with compression ratios in the range of 30:1, requiring a high-pressure external fuel gas compressor. With advanced system developments, aeroderivatives are approaching 45% simple cycle efficiencies (LHV).

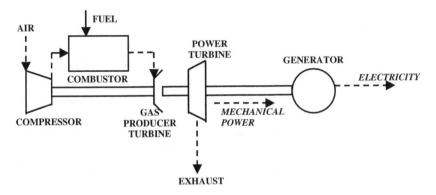

Figure 6.1 Basic components of a simple cycle gas turbine.

Industrial or frame gas turbines were developed exclusively for stationary power generation and are available in the 1 to 250 MW capacity range. They are generally less expensive and more rugged, can operate longer between overhauls, and are more suited for continuous base load operation with longer inspection and maintenance intervals than aeroderivative turbines. However, they are less efficient and much heavier. Industrial gas turbines generally have more modest compression ratios (up to 16:1) and often do not require an external fuel gas compressor. Industrial gas turbines are approaching simple cycle efficiencies of approximately 40% (LHV) and combined-cycle efficiencies of 60%.

Research into hybrid gas turbine-fuel cell systems holds promise of increasing efficiency beyond 60% (LHV) while retaining extremely low emissions.

Gas turbines operate on natural gas, synthetic gas, landfill gas, and fuel oils. Plants are often designed to operate on gaseous fuel with a stored liquid fuel for backup, so as to obtain the less-expensive interruptible rate for natural gas. Dual fuel combustion capability is a purchase option on many gas turbines. The turbine is ideally suited as an integral component in the IGCC train. Further, modern gas turbines have proven to be reliable power generators, given proper maintenance. Time to overhaul is typically 25,000 to 50,000 hours. Gas turbines can be selected to match the electric demand of most end users. Gas turbines burning gaseous fuels feature lean premixed burners (also called dry low-NO_X combustors) that produce NO_X emissions below 25 ppm, with laboratory data down to 9 ppm, and simultaneous low CO emissions acceptable to regulators and safety personnel in the 50 to 10 ppm range. Further reductions in NO_X and CO can be achieved by use of selective catalytic reduction (SCR) or catalytic combustion. Many gas turbines sited in locales with extremely stringent emission regulations use SCR after-treatment to achieve single-digit (below 9 ppm) NO_X emissions. Because gas turbines reduce power output by reducing combustion temperature, efficiency at part load can be substantially below that of full power efficiency.

FUEL REQUIREMENTS

All gas turbines intended for service as stationary power generators in the U.S. are available with combustors equipped to handle natural gas fuel. A typical range of heating values of gaseous fuels acceptable to gas turbines is 900 to 1100 Btu/scf, which covers the range of pipeline-quality natural gas. Clean liquid fuels are also suitable for use in gas turbines.

Special combustors developed by some gas turbine manufacturers are capable of handling cleaned gasified solid and liquid fuels. Burners have been developed for medium Btu fuel (in the 400 to 500 Btu/scf range), which is produced with oxygen-blown gasifiers, and for low Btu fuel (90 to 125 Btu/scf), which is produced by air-blown gasifiers. These burners for gasified fuels were principally developed for large gas turbines and are not found on small gas turbines. Contaminants in fuel such as ash, alkalis (sodium and potassium), and sulfur result in alkali sulfate deposits, which impede flow, degrade performance, and cause corrosion in the turbine hot section. Fuels are permitted to have only low levels of specified contaminants in them (typically less than 10 ppm total alkalis, and single-digit ppm of sulfur).

Liquid fuels require their own pumps, flow control, nozzles, and mixing systems. Many gas turbines are available with either gas or liquid firing capability. In general these gas turbines can be converted from one fuel to another quickly. Several gas turbines are equipped for dual firing and can switch fuels with minimal or no interruption.

Lean burn/dry low-NO_X combustors can generate NO_X emissions levels as low as 9 ppm (at 15% O_2), while those with liquid fuel combustors have NO_X emissions limited to approximately 25 ppm (at 15% O_2). There is no substantial difference in general performance with either fuel. However, the different heats of combustion result in slightly higher mass flows through the expansion turbine when liquid fuels are used, and thus a very small increase in power and efficiency performance is obtained. Also, the fuel pump work with liquid fuel is less than with the fuel gas booster compressor, thereby further increasing net performance with liquid fuels.

Gas turbines operate with combustors at pressure levels from 75 to 350 psi. Pipeline pressure of natural gas is always above these levels, and hence the pressure is usually let down during city gate metering and subsequent flow through the distribution piping system and customer metering. Often a fuel gas booster compressor is required to ensure that fuel pressure is adequate for the gas turbine flow control and combustion systems. The cost of such fuel gas booster compressors adds to the installation capital cost. Normally, redundant booster compressors are employed to ensure reliable operation, because without adequate fuel pressure a gas turbine will not operate. Liquid-fueled gas turbines use pumps to deliver the fuel to combustors. Gasification technologies generally operate within or close to the pressure levels of gas turbine combustors, and hence for some processes the syngas from a gasifier can be fired directly into a gas turbine to produce electricity (integrated gasification combined cycle).

Use of Coal-Derived Liquid Fuel

Methanol is a clean-burning liquid that can be used to power electricity-generating turbines as well as a fuel for automobiles and other vehicles. It can also be a valuable feedstock for a variety of chemicals (e.g., in the manufacture of acetic anhydride and dimethyl terephthalate). Although most methanol today is produced from natural gas, the ability to derive it from producer gas is both feasible and appears close to being practical. Any of the gasification technologies that are capable of producing medium- to high-quality syngas can be retrofitted with conventional technologies for methanol (or ethanol) synthesis.

One company that claims to have progressed significantly in this area is Eastman at its Kingsport, TN, facility. Eastman is being funded under the Clean Coal Technology Program. The DOE's industrial partner in the project is the Air Products Liquid Phase Conversion Company, L.P., a joint venture between Air Products and Eastman. As part of the Clean Coal Technology Program, the company has designed and constructed a liquid-phase methanol (LPMEOH) process demonstration facility as an add-on to the Kingsport complex. The most recent information we have is that the demonstration facility produces 260 tons per day of methanol.

Liquid-phase methanol synthesis technology is an outgrowth of more than a decade of federally sponsored research. The unique feature of the Eastman technology is that it converts coal-derived synthesis gas to methanol in a single vessel containing catalyst particles suspended in mineral oil. More conventional technology, by contrast, sends the synthesis gas through a fixed bed of dry catalyst particles. The liquid-phase process is much better suited for directly processing the gases produced by current coal gasifiers. The liquid-phase technology offers other advantages over conventional technologies including greater stability and heat dissipation in the conversion process, reliable on/off operation, and the ability to use the product methanol directly as a fuel without further upgrading.

In future commercial facilities, advanced coal-to-methanol processes may be a cost-enhancing option for coal gasification-based power plants. As already noted, other demonstration projects in the Clean Coal Technology Program are showing the attractiveness of the IGCC technology. In this approach, gas made from coal is burned in a gas turbine to generate electricity, and the exhaust heat is used in a conventional steam generator. In future facilities of this type, producing methanol as a co-product during times of low-electricity demand (at night, for example) would allow the gasifiers to operate at steady, peak performance.

Methanol contains no sulfur and produces very little nitrogen oxide pollutants when burned, making it a very clean combustion fuel. At a power generating facility, it could be used as a supplemental fuel for gas turbines to meet peak electricity generation requirements, or it could be sold "over the fence" to commercial fuel and chemical companies. A commercial-scale power facility might generate 200 to 350 MW of electricity, while also producing 150 to 1000 tons per day of methanol.

The production of dimethyl ether (DME) as a mixed co-product with methanol also appears highly feasible. DME has several commercial uses. In a storable blend with methanol, the mixture can be used as a peaking fuel in gasification-based electric power generating facilities. DME can also be used to increase the vapor pressure of methanol, making it suitable for use as a diesel engine fuel. Blends of methanol and DME can be used as chemical feedstocks for synthesizing chemicals, including new oxygenated fuel additives.

MARKET TRENDS

Of the next 1000 power plants to be built in the U.S., as many as 900 of them are expected to use natural gas turbines. A report to the National Renewable Energy Laboratory indicates growing numbers of gas turbines in the 5 to 40 MW size range are being placed by utilities at substations to provide incremental capacity and grid support. This defers the need for transmission and distribution (T&D) system expansion, can provide temporary peaking capacity within constrained areas, or can be used for system power factor correction and voltage support, thereby reducing costs for both customers and the utility system. Turbine manufacturers and packagers now offer mobile turbine generator units in this size range that can be used in one location during a period of peak demand and then moved by truck to another location. While the use of small gas turbines is expected to grow in both grid support and customer peak-shaving applications, the most prevalent on-site generation application for small gas tur- bines has traditionally been CHP. The general industry liter- ature indicates that this trend is likely to continue. The economics of small gas turbines in on-site generation appli- cations often depend on effective use of the thermal energy contained in the exhaust gas, which generally represents 60 to 70% of the inlet fuel energy.

The number of U.S. electric utilities offering distributed generation (DG) and CHP services/products has more than doubled in the past 18 months and now stands at almost 40%, according to one Web site source (http://www.spratley.com). There is a market for at least 20 GW of new distributed electric generation in the U.S. alone. And even larger markets exist internationally, where grids are poor or the difference between gas and electric costs is higher than in the U.S. In addition to the emerging distributed generation market, there are 50 GW of demand for CHP installations in the U.S.

Another source (http://www.distributed-generation.com/ market_forecasts.htm) estimates there is at least 34 GW of installed U.S. DG capacity. The Gas Research Institute esti- mates installed DG capacity in the U.S. (of all unit sizes, including those below 300 kW) as 75 GW (this is an estimate as of 2000). So the installed base is likely in the 34 to 75 GW range. Most of the installed base are units less than 2.0 MW in size. Nearly all of the installed base is diesel reciprocating engines. According to Diesel and Gas Turbine's annual survey of unit orders (over 1 MW), turbines provide over 90% of new capacity from engines or turbines. However, in the DG size range (under 50 MW), most turbine growth has been for central station peakers (not a DG application). Smaller, DG baseloaded turbines (mostly CHP) have actually declined, from about 160 MW in 1998–99 and 1999–2000 to about 20 MW in 2000–01.

For comparative purposes, in the past 2 years, mid-sized reciprocating engines (2 to 3.5 MW) provided explosive growth in new reciprocating engine capacity. Capacity from orders of these units quadrupled from 1998 to 2001; during the same time period smaller (1 to 2 MW) units grew by 40%. The 1 to

TABLE 6.2   Market Sectors and Applications

Market Sector	Market Application	Capacity (MW)	Number of DG Units (thousands)
Industrial	Continuous power	900	3
	CHP	9,000	35
	Peak shaving	500	2
	Standby/emergency	5,000	100
	Premium power	8,000	30
Commercial	Standby/emergency	15,000	300
	Premium power	30,000	600
Residential	Fuel cells	3,500	500

2 and 2 to 3.5 MW reciprocating units are mostly peaking/standby units fueled by diesel, with some continuous duty natural gas fired applications. Of these, many can be attributed to data centers, as well as backup power for larger commercial and industrial facilities.

The potential market size forecast for DG varies by application. Table 6.2 indicates the approximate size of some key U.S. markets, for a few of the customer applications. Any forecast depends greatly on the expected DG capital cost, future fuel cost, and the technology's energy efficiency. All three variables differ greatly by DG technology, making for different market niche applications. These estimates were made by identifying specific instances where DG economics can beat grid costs. This does not mean actual sales will be this high; rather, it suggests the size of the economic potential market within which DG developers are operating. As all these variables are changing rapidly, only estimates of sales in the U.S. for 2001–2010 are provided here. Any market forecast depends on how quickly regulatory barriers and other development issues are addressed. Nonetheless, these numbers provide an indication of the size of the potential market.

In the U.S., there is an installed base of mostly smaller reciprocating engines of about 50 GW. Recent annual sales have been 20 MW of DG turbines and 7,500 MW of reciprocating engines. With microturbines, fuel cells, and other DG renewables just now coming online, less than 100 MW of capacity annually of these technologies has been sold in recent years. Nonetheless there is the potential to double DG's installed capacity by adding as much as 72 GW by 2010. It should be noted that the portfolio of DG technologies includes reciprocating engines, microturbines, combustion gas turbines (including miniturbines), fuel cells, photovoltaics, and wind turbines. Each technology has varying characteristics and emission levels. Table 6.3 summarizes some of the key specifications for each technology.

Microturbines are an emerging class of small-scale distributed power generation in the 30 to 400 kW size range. The basic technology used in microturbines is derived from aircraft auxiliary power systems, diesel engine turbochargers, and automotive designs. A number of companies are currently field-testing demonstration units, and several commercial units are available for purchase. These machines consist of a compressor, combustor, turbine, and generator. The compressors and turbines are typically radial-flow designs and resemble automotive engine turbochargers. Most designs are single-shaft and use a high-speed permanent magnet generator, producing variable voltage, variable frequency alternating current (AC) power. Most microturbine units are designed for continuous-duty operation and are recuperated to obtain higher electric efficiencies. As this market grows and applications emerge, coal gasification will likely capture niche opportunities by providing coal-derived liquid fuels such as methanol, ethanol, or fuel blends to fuel these machines.

Combustion turbines range in size from simple cycle units starting at about 1 MW to several hundred MW when configured as a combined cycle power plant. Units from 1 to 15 MW are generally referred to as industrial turbines (or sometimes as miniturbines), which differentiates them from both larger utility grade turbines and smaller microturbines. Units smaller than 1 MW exist, but few have been installed in the U.S. Industrial turbines are currently available from numerous manufacturers. As noted earlier, historically, they were developed as aeroderivatives, spawned from engines used for jet propulsion. Some, however, are designed specifically for stationary power generation or compression applications in

TABLE 6.3   Key Specifications for Competing Technologies

Technology	Recip Engine: Diesel	Recip Engine: NG	Microturbine	Combustion Gas Turbine	Fuel Cell
Size	30kW–6+MW	30kW–6+MW	30–400kW	0.5–30+MW	100–3000kW
Installed Cost ($/kW)[1]	600–1000	700–1,200	1200–1700	400–900	4,000–5,000
Elec. Efficiency (LHV)	30–43%	30–42%	14–30%	21–40%	36–50%
Overall Efficiency[2]	~80–85%	~80–85%	~80–85%	~80–90%	~80–85%
Total Maintenance Costs[3] ($/kWh)	0.005–0.015	0.007–0.020	0.008–0.015	0.004–0.010	0.0019–0.0153
Footprint (sqft/kW)	0.22–0.31	0.28–0.37	0.15–0.35	0.02–0.61	.9
Emissions (gm/bhp-hr unless otherwise noted)	NO_x: 7–9 CO: 0.3–0.7	NO_x: 0.7–13 CO: 1–2	NO_x: 9–50ppm CO: 9–50ppm	NO_x: <9–50ppm CO:<15–50ppm	NO_x: <0.02 CO: <0.01

[1] Cost varies significantly based on siting and interconnection requirements, as well as unit size and configuration.

[2] Assuming CHP.

[3] Fuel costs vary significantly by region, and the numbers in this row assume minimum fuel cost. For a specific application, contact manufacturers of specific equipment to get an idea of fuel consumption. Contact local gas, diesel, or other fuel supplier (or see Department of Energy databases) to get an estimate of what fuel will cost.

the oil and gas industries. Multiple stages are typical and along with axial blading differentiate these turbines from the smaller microturbines described above. Combustion turbines have relatively low installation costs, low emissions, and infrequent maintenance requirements. However, their low electric efficiency has limited turbines to primarily peaking unit and CHP applications. Cogeneration DG installations are particularly advantageous when a continuous supply of steam or hot water is desired.

R&D NEEDS

Improved Operational Performance

Many operational conditions affect the propensity to fail in a gas turbine. Frequent starts and stops incur damage from thermal cycling, which accelerates mechanical failure. Use of liquid fuels, especially heavy fuels and fuels with impurities, radiates heat to the combustor walls significantly more intensely than occurs with clean, gaseous fuels, thereby over-heating the combustor and transition piece walls. On the other hand, steady operation on clean fuels can permit gas turbines to operate for a year without need for shutdown. Estimated availability of gas turbines operating on clean gaseous fuels like natural gas is in excess of 95%. Use of distillate fuels and other liquid fuels, especially those with contaminants (alkalis, sulfur, and ash), requires more frequent shutdowns for preventative maintenance, which reduces availability. These factors suggest the need for more robust machines that are capable of accepting "less than pure" forms of liquid and gaseous fuels, as may be derived from coal gasification. The nature of R&D efforts should focus both on machine performance integrity improvements and on the development of economical syngas cleaning and liquid fuels refining.

Improved Efficiencies

Improvement to gas turbine efficiencies is a general goal of manufacturers. It further enhances the attractiveness of incorporating gasification into integrated systems since energy in the form of steam is needed to operate gasifiers. There are several areas of R&D that focus on efficiency improvements.

Gas turbine efficiency and power increases with increasing firing temperature. Increased power results in lower cost per kW, provided that the cost of the technology needed to increase the firing temperature is economic, and current high levels of system durability, reliability, availability, and maintainability (DRAM) are not compromised. Manufacturers are pursuing advances in several critical components to accommodate higher firing temperatures. The most critical limitations on firing temperature are embodied in the combined thermal and mechanical design of the gas turbine blades and vanes. The design and manufacturing process for gas turbine blades and vanes involves economic selection of materials for a combination of high strength at elevated temperature and manufacturability, and heat transfer analysis of the component with internal cooling and surface thermal protection coating so that the creep limit of the material is not exceeded during operation. Firing temperature is also limited by design, as higher firing temperature invariably results in increased NO_X emission rates. The mechanical integrity of hot section components (i.e., combustor, transition pieces, and turbine section blades and vanes) is critical. As firing temperature increases, the heat flux on the combustor and transition piece to the power turbine increases. The transition piece is the component that ducts the combustor exit gases into the first stage vanes. The increase in temperature of such structural components is a major concern in achieving reliable operation with increased firing temperature. While the alloys used for these components have advanced little in recent times, the manufacturing process has improved significantly. As firing temperature increases, the heat load on the combustor walls also increases rapidly, adding the combustor to the list of critical hot section components. Because of the effect of temperature on the life of hot section parts, gas turbines usually are rated for continuous duty (long life) and for emergency or peaking duty.

Compressor and turbine blade and vane aerodynamic efficiencies are also important in obtaining the highest system efficiency. Continuous research focuses on improved shapes of blades and vanes. These advances usually begin commercial service on new gas turbine models entering the market. At later dates, these improvements in component efficiency are incorporated into existing machines as upgrades. Three-dimensional fluid dynamics computational analysis has helped to increase the efficiency of these aerodynamic components, thereby reducing aerodynamic losses and increasing overall machine efficiency. The newest advancements include three-dimensional blade and vane profiling for the purpose of reducing losses attributable to secondary flows occurring in boundary layers. Such blades and vanes have tailored radial contouring.

Gas turbine efficiency and power increase with increasing turbine inlet temperature. However, turbine blade and vane materials have well-defined temperature limitations consistent with long life and safe operation. In order to increase the turbine inlet temperature without overheating the materials, manufacturers have introduced internal cooling into the design of blades and vanes. Internal cooling takes high-pressure air from the compressor and diverts a small, engineered percentage to the expansion turbine section, where compressor air is used to cool the turbine disks holding the blades. The air passes through a network of precise holes and flow passages inside the blades and vanes, cooling these components, and then passes into the main flow at the downstream end of the blade or vane. The internal cooling creates a temperature gradient in the blade or vane material and reduces both the average and local maximum material temperature. Detailed knowledge of heat transfer coefficients over surfaces and within flow passages allows for judicious design of blades and vanes as heat exchangers. This design allows the temperature of the gas expanding in the turbine to be increased while keeping the material at or below its safe long-life design temperature. The use of compressor air for cooling does reduce the mass flow through that stage of the turbine and consequently slightly reduces power. When the tradeoff is complete, appreciable increases in turbine gas flow temperature result in substantial gains in both power and efficiency. Turbine blade and vane cooling was developed for aircraft gas turbines, where the additional power is vital to increasing an airplane's top speed and acceleration. Over time, cooling became available for stationary gas turbines and was disseminated through engineering technical papers and the migration of aircraft gas turbine engineers to stationary products. Cooling is now appearing in smaller gas turbines, where product economics precluded the level of investment in advanced technology development that is economic for larger machines.

Another technology used to allow increased temperature of the expansion gas without overheating blade and vane materials is surface coating of the materials that are exposed to high temperature and have some degree of internal cooling. These surface barrier coatings are generally ceramic and provide both corrosion resistance and thermal protection. The coatings are capable of sustained life at elevated temperature. With internal cooling, a heat flux through the material exists. This heat flux results in temperature drops through the coating on the surfaces and in the surface (gas) boundary layer. Unfortunately, these coatings have been limited to thin layers, as their differential thermal expansion is quite different from that of the substrate metal, and thick coatings tend to separate from the base. However, because adequate protection is afforded with thin sections, the use of such thermal barrier coating is an established practice.

Ceramic materials that retain structural integrity to temperatures in the 2100 to 2400°F range have been the subject of research and development for many years. Researchers have in fact created small radial inflow turbines from structural ceramic material for possible use in automotive gas turbines. These experimental units have shown favorable properties in laboratory tests. However, several practical considerations pose potential stumbling blocks to their use in commercial systems, such as coefficients of expansion that are substantially different from those of the metals used in gas turbine construction. One may expect to find ceramic materials in use in industrial gas turbines in the future, first on vanes and then on blades in the expansion turbines. The benefit of the use of ceramics is that vane and blade cooling need not be used. This reduces the extra cost of building vanes and blades with complex internal structures, and eliminates the need for some of the compressor air to be used for cooling. With greater air flowing through the combustor, more power can be generated in the first turbine stages, with resulting gains in both power and efficiency. Experimental industrial gas turbines with ceramic blades and vanes have operated for over 9000 hours in laboratory testing, and a unit in the field is currently undergoing evaluation. However, commercial availability is uncertain, as these ceramic components are brittle and subject to sudden catastrophic damage by ingestion and impact by foreign objects.

FUEL CELL TECHNOLOGY DEVELOPMENT STATUS

Fuel cells are electrochemical systems that convert the energy of a fuel directly into electric power. The design of a fuel cell is based on the key components: an anode, to which the fuel is supplied; a cathode, to which the oxidant is supplied; and an electrolyte, which permits the flow of ions (but no electrons and reactants) from anode to cathode. The net chemical reaction is exactly the same as if the fuel was burned, but by spatially separating the reactants, the fuel cell intercepts the stream of electrons that spontaneously flow from the reducer (fuel) to the oxidant (oxygen) and diverts it for use in an external circuit.

The main difference between a fuel cell and a battery is that the fuel and oxidants are not integral parts of the fuel cell, but instead are supplied as needed to provide power to an external load, while the waste products are continuously removed. Where hydrogen is supplied as the fuel to the anode and oxygen to the cathode, this waste product is only water.

A single cell produces a voltage of about 1 V. To obtain higher voltages the cells are connected in series to form a "stack." Furthermore, heat rejected in the process can be used for different on-site thermal consumption, which makes the fuel cells suitable for combined heat and power generation for buildings and industry.

Like combustion systems, fuel cells can use oil, natural gas, coal, and methanol. These fuels, however, have to be pre-processed into a suitable hydrogen-rich form. Fuel cells can also use hydrogen obtained by electrolysis of water using stand-by electricity from photovoltaics, wind energy, or hydrogen-rich or refined syngas produced from coal gasification. A demonstration plant for this technical chain has been built in Neunburg vorm Walde, Germany, by Solar-Wasserstoff Bayern GmbH.

The overall advantages of fuel cells are the low environmental impact, which is one to two orders of magnitude lower than in conventional systems, good part load behavior, easy operation, and low maintenance since no rotating parts are needed. The main disadvantages at the moment are the very high costs and the lack of demonstrated reliability.

Different fuel cell types exist. They operate at different temperatures and are generally distinguished by their electrolytes. The status of development differs widely for each type. Table 6.4 provides a comparison of the major types of fuel cells currently under development.

Different fuel cells are at different stages of development. The high-temperature fuel cells (molten carbonate and solid oxide fuel cells, or MCFCs and SOFCs) are at an early development stage. Basic research is still needed to solve material problems such as corrosion in MCFC or decreasing the operating temperature for SOFC. Market introduction for MCFC is not expected before 2005 and for SOFC not before 2010. Phosphoric acid fuel cells (PAFC)s have reached the stage of demonstration. They have been commercially available since 1990 and so far sixty 200 kW units have been sold worldwide. The latest research topic is the use of stationary polymer electrolyte fuel cell (PEMFC) in domestic applications. This research field can be seen as a by-product of PEMFC activities for mobile application. The oldest type of fuel cells, the alkaline fuel cell (AFC), is on the market now. The restricted use of pure hydrogen and oxygen as fuel has limited fuel cell application possibilities to space and military uses.

TABLE 6.4 Comparison of Major Types of Fuel Cells

Fuel Cell Type	Electrolyte	Temp., °C	Efficiency, %	Advantages	Disadvantages	Operation Field
AFC	Alkaline	80	60	High current and power densities, high efficiencies	CO_2 intolerance	Space power, military
PEMFC	Polymer	80–110	60	High current and power densities, long operating life	CO_2 intolerance, water management, noble catalyst	Transportation, cogeneration
PAFC	Phosphoric acid	200	40	Technology well advanced	Relatively low efficiency, limited lifetime, noble catalyst	CHP
MCFC	Molten carbonate	650	48–56	High efficiency, internal fuel processing, high-grade waste heat	Electrolyte instability, short operating life, CO_2 recycling	Power production, cogeneration
SOFC	Solid oxide ceramic	1000	55–65	High efficiency, internal fuel processing, high-grade waste heat, long operating life	High operating temperature, relatively low ionic conductivity	Power production, cogeneration

Operating temperature is a critical parameter that determines the potential uses of each type of fuel cell. For example, alkaline, direct methanol (DMFC), and solid polymer fuel cells (SPFC) have potential applications in transport because they do not produce much heat (which otherwise would have to be eliminated by some cooling device) and, as a result of this, have a very short start-up period (a few minutes). In contrast, phosphoric acid, molten carbonate, and solid oxide fuel cells producing high-temperature heat are more complex to run and are a better fit for stationary applications like power generation or CHP. Low-temperature fuel cells (<200°C) are very dependent on expensive catalysts (usually made of platinum or other noble metals) to achieve adequate reaction rates, and need to use pure hydrogen as a fuel, while high-temperature fuel cells can avoid using a catalyst and can use natural gas or synthetic gas (from coal gasification) as fuel. In fact the presence of high temperatures also opens the possibility of internal reforming of the fuel (within the fuel cell itself), although most of the experience to date is with external or partially external reforming, using heat from the fuel cell (see Patel, 1995).

A typical problem to fuel cells operating at low temperatures comes from the catalyst, which can be damaged (or "poisoned") by the presence of CO or CO_2 and needs to be replaced: AFC and PEMFC are rather intolerant to CO_2 and CO, while PAFC is moderately tolerant to CO and MCFC and SOFC are fully tolerant to CO.

Depending on the operating temperature and the complexity of the device some fuel cells types are better fit than others to small-scale residential applications (typically in the range from 30 kWe to 1 MWe), although all types of FCs have to be commercially viable at the small scale before they can be viable at the large scale.

Possible applications of fuel cell includes producing small scale electricity only (low temperature FCs) or heat and power for houses, residential buildings, hotels, hospitals, sport facilities, and shopping centers. Other applications in urban situations, on a larger scale (up to 30 MWe), could be for distributed generation of heat and power or of power only.

A typical cell generates a voltage of around 0.7 to 0.8 volts per cell and power outputs of a few tens or hundreds of watts. In order to achieve a significant output, cells have to be assembled in modules or stacks and electrically connected in series or in parallel. Different types of cells exist, according to the electrolyte used, and each type has a characteristic operating temperature (OT):

- Alkaline fuel cells use a solution of potassium hydroxide as an electrolyte and have an OT of 30 to 200°C (Patel (1995).
- Direct methanol fuel cells use sulfuric acid or a polymer membrane as an electrolyte and have an OT of 80 to 130°C.
- Phosphoric acid fuel cells use phosphoric acid as an electrolyte and have an OT of 190 to 210°C.
- Solid polymer fuel cells use sulfonic acid group in polymer as an electrolyte and have an OT of 50 to 80°C.
- Molten carbonate fuel cells use a mixture of lithium and potassium carbonate as an electrolyte and have an OT of 630 to 650°C.
- Solid oxide fuel cells use zirconium oxide stabilized with yttrium as an electrolyte and have an OT of 850 to 1000°C.

It is generally acknowledged that, due to their extremely low emissions of SO_2 and NO_X (< 5 ppm for NO_X), fuel cells have an interesting market potential especially in situations where environmental standards are very stringent, i.e., within cities or nonattainment areas, for repowering of old plants, or to supply power and heat closer to demand centers, avoiding the grid power losses that would be unavoidable if plants were built in distant sites. However, this is particularly true for small applications, rather than for larger size plants, where the competition afforded by IGCC is very strong both on the cost side and on the energy efficiency side (electric efficiency close to 60%), and very high levels of emissions abatement can be reached using appropriate technology. CO_2 emission, furthermore, in most fuel cell types is still a problem that must be addressed.

The advantage of modularity that fuel cell stacks have does not imply necessarily that a balance of plant for a given size unit can be replicated n times to obtain a plant with capacity n times greater and optimal design. The most economical design for that size of plant may be entirely different, and hence, the plant must always be carefully designed. Fuel cells do leave more room for standardization and modularity of units than most technologies.

For large-scale power generation, MCFCs and SOFCs are expected to make some impact in the long-term horizon (2010–2020), as this technology is at the development stage (a 2 MW MCFC plant was built by Southern California Edison and Energy Research Corporation and started in the spring of 1996; another of the same size was planned by ECN but the project has been abandoned and ECN is now concentrating on internal reforming in small - 10 kW - fuel cells). Hybrid versions (fuel cells + coal gasification plants) could be potentially more efficient and less environmentally harmful than current state-of-the-art IGCC plants, but a number of system-level problems still need to be solved.

For smaller scale, multi-MW distributed power systems a market could potentially open in the shorter term; opportunities for plants under 1 MW are only seen for SOFCs but are hard to envision for MCFCs due to their inherent complexity.

Thus far the only options that have reached a commercial scale are PAFCs and PEMFCs, but the latter, although potentially interesting also for power generation in stationary settings, are being developed almost exclusively for transport applications (no PEMFC plant is currently operating for stationary application). Compared to other fuel cell types, PAFCs have relatively low efficiency rates, unless pure hydrogen is used as feed, but the technology is strengthening due to the experience being accumulated in a number of countries (based on about 50 MW of installed capacity, worldwide). Availability factors for these plants (mostly < 1 MW in size) already exceed 95%, and problems come either from corrosion in the primary cell stack coolant circuit or from other plant components (pumps, filters, fan drives). In Europe at least 11 PAFC plants of about 200 kW capacity (of which 10 have operated for over

4 years), and 4 plants of 50 kW capacity, have been installed. In Italy a 1 MW PAFC plant, built in collaboration between Ansaldo and AEM, has been working for over 2 years. Another 200 kW plant has been built in Bologna for the municipal company ACOSER. Although there are these semi-commercial successes, the development of a fully commercial market for this fuel cell type in the small-/medium-scale cogeneration sector is likely to take at least another 5 years. Costs need to be abated further.

Worldwide over 150 demonstration plants have been installed. These represent around 40 to 50 MW of electrical generating capacity. Nearly 75% is installed in Japan, over 15% in North America, and 9% in Europe. The U.S.-based International Fuel Cells (IFC) and its partner Toshiba are responsible for producing over 70%, Fuji over 25%, and Mitsubishi about 2% (WFCC analysis). These phosphoric acid systems are operating in "real world" commercial situations and have clearly demonstrated their suitability for on-site cogeneration.

Typical service life and stack costs are reported in Table 6.5.

Although an FC produces electricity, an FC power system requires the integration of many components beyond the fuel cell stack itself, for the FC will produce only DC power and utilize only processed fuel. Various system components are incorporated into a power system to allow operation with conventional fuels, to tie into the AC power grid, and often,

TABLE 6.5   Life Demonstration and Stack Costs

Fuel Cell	Catalyst	OT, °C	Carbon Tolerance	Life Demonstrated, hr	Stack Cost, $/kW
SPFC	Pt	25–85	0.01% CO	57,000	730
Alkaline	Ni/Pt	65–220	None	10,000	220
PAFC	Pt	150–205	1.5% CO	10,000	400
MCFC	Ni	600–650	Good	3000	300
DMFC	PT, Pt/Ru	60–300	0.01% CO	—	750
SOFC	Ni or Co	800–1000	Good	1600	750

to utilize rejected heat to achieve high efficiency. In a rudimentary form, fuel cell power systems consist of a fuel processor, fuel cell power section, power conditioner, and potentially a cogeneration or bottoming cycle to utilize the rejected heat. System design examples for present-day and future applications are presented in the following sections, along with a discussion of research and development areas that are required for the future system designs to be developed.

Integrated Gasification Fuel Cell Power Systems Requirements

Figure 6.2 provides a simplified process schematic of an FC power plant. Syngas from the gasification plant cleanup system is first cleaned further and moisturized. With a carbonate FC, the moisturized syngas is fed to the anode side of the FC, where methane is internally reformed and CO is shifted to CO_2 and H_2. Spent fuel exits the anode and is further oxidized in the anode exhaust oxidizer to supply oxygen and CO_2 to the cathode. Resulting actions in the FC anode and cathode produce DC output, which can then be converted to AC. The cathode exhaust supplies heat to the fuel cleanup, the steam boiler, and cogen system as it is vented from the plant.

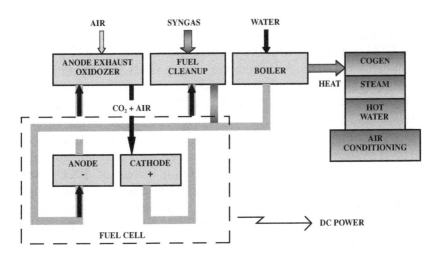

Figure 6.2 Simplified process of a fuel cell power plant.

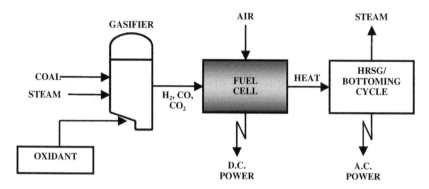

Figure 6.3 Simplified block diagram of IGFC system.

Figure 6.3 provides a simplified block diagram of a fuel cell power plant system. Gasification is used to convert the solid fuel to gas, which is processed to remove sulfur compounds, tars, particulates, and trace contaminants. The clean gas is then converted to electricity in the FC. Waste heat from the FC is used to generate steam, which can be used to run the gasification process and to generate additional power in the bottoming cycle.

Fuel Cell Energy Inc. has reported that it is planning to build and test a 2 MW FC power plant for operation of coal-derived gas. The proposed plant is based on Direct Fuel Cell (DFC™) technology and is part of a Clean Coal V IGCC project supported by DOE. A British Gas Lurgi (BGL) slagging fixed-bed gasification system with cold gas cleanup is a part of a 400 MW IGCC power plant that will provide a fuel gas slip stream to the FC.

The above studies all substantiate the feasibility of integrated systems, and indeed, designs are well past the conceptual stage. There are, however, major areas of R&D that require support in scaling up fuel cells and achieving demonstrated integration with gasification to place the technology on an economical level. These areas are discussed in the next sections.

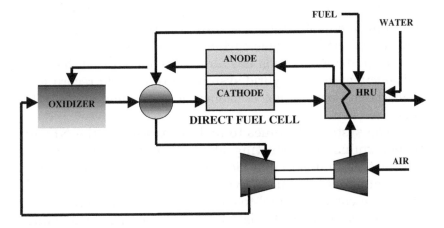

Figure 6.4   High-efficiency hybrid FC/Turbine power cycle.

Integrated Gasification Fuel Cell Hybrid Power Systems Requirements

Hybrid systems that incorporate gas turbines build upon the outstanding performance of the fuel cell by utilizing the exhausted FC heat. Hybrid FC/turbine systems have shown that LHV efficiencies of 70% can be achieved on natural gas (see Ghezel-Ayagh et al. (1999, 2000, 2001)). The proposed configuration uses a gas turbine as a bottoming cycle to the fuel cell. Figure 6.4 illustrates the system configuration. There are no theoretical barriers to applying the same system configuration to coal gas systems.

The first SOFC/gas turbine hybrid system was delivered to the National Fuel Cell Research Center (NFCRC) in 2000 for operation and testing on behalf of Southern California Edison and in cooperation with Siemens Westinghouse Power Corporation. This is a 220 kW hybrid SOFC/microturbine system. The hybrid system includes a pressurized Siemens Westinghouse SOFC module integrated with a microturbine/generator supplied by Ingersoll-Rand Energy Systems (formerly Northern Research and Engineering Corp.). The system has a total output of 220 kW, with an output of approximately

180 kW from the SOFC and approximately 40 kW from the microturbine generator. This system is the first-ever demonstration of the SOFC/gas turbine hybrid concept. This proof of concept demonstration has already demonstrated the high-efficiency feature of hybrid systems with a world record fuel-to-electricity conversion efficiency of approximately 53% for this size class.

The system continues to undergo testing at the NFCRC to determine its operating characteristics and operating parameters, and gain experience for the design of prototypes and commercial products. Eventually, such SOFC/GT hybrids should be capable of fuel-to-electricity conversion efficiencies of 60 to 70%. If an SOFC is pressurized, an increased voltage results, leading to improved performance. For example, operation at 3 atmospheres increases the power output by approximately 10%. However, this improved performance alone may not justify the energy required for pressurization. What will justify the expense of pressurization is the ability to integrate the SOFC with a gas turbine that needs a hot pressurized gas flow to operate. Since the SOFC stack operates at 1000°C, it produces a high-temperature exhaust gas. If operated at elevated pressure the exhaust becomes a hot pressurized gas flow that can be used to drive a turbine. If an SOFC and a gas turbine are integrated carefully, the pressurized air needed by the SOFC can be provided by the gas turbine's compressor, the SOFC can act as the system combustor, and the exhaust from the SOFC can drive the compressor and a separate generator. This yields a dry (no steam) hybrid-cycle power system of unprecedented electrical generation efficiency. As of summer 2002, the system was tested over four separate periods of time for an excess of 1500 hours.

Figure 6.5 provides a detailed process flow diagram of the system. During normal operation air enters the compressor and is compressed to ~ 3 atmospheres. This compressed air passes through the recuperator, where it is preheated and then enters the SOFC. Pressurized fuel from the fuel pump also enters the SOFC, and the electrochemical reactions take place along the cells. The hot pressurized exhaust leaves the SOFC and goes directly to the expander section of the gas

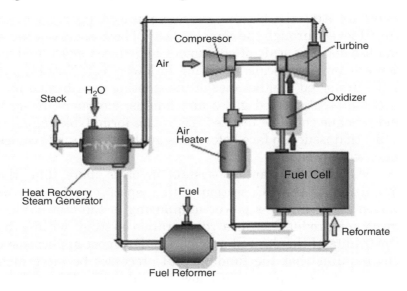

Figure 6.5 Simplified diagram of hybrid system under development.

turbine, which drives both the compressor and the generator. The gases from the expander pass into the recuperator and then are exhausted. At ~ 400°F the exhaust is hot enough to make hot water. Electric power is thus generated by the SOFC (DC) and the gas turbine generator (AC) using the same fuel/air flow. Analysis indicates that with such SOFC/GT hybrids an electrical efficiency of 58% can be achieved at power plant capacities as low as 250 kW, and 60% as low as 1 MW using simple small gas turbines. At the 2 to 3 MW capacity level with larger, more sophisticated gas turbines, analysis indicates that electrical efficiencies of 70% or more are possible.

System Configurations and Costs

In the United States, in the commercial sector, the most attractive application for near-term fuel cell technology is on-site cogeneration. In this application, the fuel cell power plant is located at the site of the end user, providing both electric and thermal energy. On-site applications in the commercial sector are attractive because the maximum energy efficiencies are achieved through the combined use of both energy outputs and because the scale of equipment introduces mass production economies of scale, resulting in lower cost. Natural gas is the preferred fuel because of the existing distribution network but coal-derived gases and liquids, methanol, bio-gas, and propane could be utilized. There are formidable commercialization issues to face before this application area becomes more widespread.

The design of an FC system involves more than the optimizing of the FC section with respect to efficiency or economics. It involves power minimizing of the cost of electricity (or product, as in a cogeneration system) within the constraints of the desired application. For most applications, this requires that the fundamental processes be integrated into an efficient plant with low capital costs. Often these objectives are conflicting, so compromises, or design decisions, must be made. In addition, project-specific objectives, such as desired fuel, emission levels, potential uses of rejected heat (electricity, steam, or heat), desired output levels, volume or weight criteria (volume/kW or weight/kW), and tolerance for risk, all influence the design of the fuel cell power system. The FC is a power generation technology that is still in the early stages of its commercial use. As a result, it is paramount to target applications that have the potential for widespread use (to attract adequate development money) with the simplest technology development (to minimize development cost). There is a strong relation between applications and the infrastructure of available fuels. Several high-value niche markets drove early fuel cell technology development. These were the use of fuel cells for on-board electric power in space vehicles and to demonstrate that fuel cells are an efficient, environmentally friendly technology for stationary on-site commercial power. As environmental regulations become more stringent for megawatt size power stations and fuel cells are scaled larger in size, there is the possibility to use the U.S.'s most plentiful, indigenous fuel, coal.

Coal covers a broad spectrum of solid fuels that complicate fuel processing, particularly cleanup. Fuel cells will find it difficult to compete economically with the high power density and relatively low-cost gas turbine power generators in this application area. Large power station operators require an alternate fuel, usually heavy oil. Also, there is the possibility of using other available fuels such as light distillates, ethanol, anaerobic digester gas, biomass, and refuse-derived fuel. However, these fuels apply to niche market applications. Fuel cell application here, if practical, will evolve from and after widespread uses. Users may require an alternate fuel, probably natural gas. Fuel processing technology from large chemical installations has been successfully transferred to small compact fuel cell units to convert pipeline natural gas, the fuel of choice for small stationary power generators.

Gas reformate is among the highest priority issues in widespread application of the technology and ultimate integration with coal gasification technologies. There are three major gas reformate requirements imposed by the various fuel cells that need addressing. These are sulfur tolerance, carbon monoxide tolerance, and carbon deposition. The activity of catalysts for steam reforming (SR) and autothermal reforming can also be affected by sulfur poisoning and coke formation. These requirements are applicable to most fuels used in fuel cell power units of present interest. There are other fuel constituents that can prove detrimental to various fuel cells. However, these appear in specific fuels and are considered beyond the scope of this review. Examples of these are halides, hydrogen chloride, and ammonia. Finally, fuel cell power unit size is a characteristic that impacts fuel processor selection. There are also discrepancies in the amounts of harmful species tolerance that fuel cell developers establish, even for similar type fuel cells. These discrepancies are due to electrode design, microstructure differences, or the way developers establish tolerance. There are some cases where the presence of certain harmful species causes immediate performance deterioration. More often, the degradation occurs over a long period of time, dependent on the developer's allowable voltage degradation rate on exposure to the specific harmful species. The developer establishes an estimated cell life based on economics. The permissible amount of the harmful constituent is then determined based on its life effects.

Sulfur. Present-formula gasolines contain approximately 300 ppm of sulfur. No. 2 fuel oil contains 2200 to 2600 ppm by weight of sulfur. Even pipeline gas contains sulfur-containing odorants (mercaptans, disulfides, or commercial odorants) for leak detection. Metal catalysts in the fuel reformer can be susceptible to sulfur poisoning, and it is important that sulfur in the fuel reformate be removed. Some researchers have advised limiting the sulfur content of the fuel in a stream reformer to less than 0.1 ppm, but noted the limit may be higher in an autothermal reformer. Sulfur poisons catalytic sites in the fuel cell also. The effect is aggravated when there are nickel- or iron-containing components, including catalysts that are sensitive to sulfur and noble metal catalysts, such as found in low-temperature cell electrodes.

Carbon Monoxide. Carbon monoxide, a fuel in high-temperature cells (MCFC and SOFC), is preferentially absorbed on noble metal catalysts that are used in low-temperature cells (PAFC and PEFC) in proportion to the hydrogen-to-CO partial pressure ratio. A particular level of carbon monoxide yields a stable performance loss. The coverage percentage is a function of temperature, and that is the sole difference between PEFC and PAFC. PEFC cell limits are < 50 ppm into the anode; major U.S. PAFC manufacturers set tolerant limits as < 1.0% into the anode; MCFC cell limits for CO and H_2O shift to H_2 and CO_2 in the cell as the H_2 is consumed by the cell reaction due to a favorable temperature level and catalyst.

Carbon Deposition. The processing of hydrocarbons always has the potential to form coke (soot). If the fuel processor is not properly designed or operated, coking is likely to occur. Carbon deposition not only represents a loss of carbon for the reaction but more importantly also results in deactivation of catalysts in the processor and the fuel cell, due to deposition at the active sites.

FUEL PROCESSING TECHNOLOGY

The generic term applied to the process of converting liquid or gaseous light hydrocarbon fuels to hydrogen and carbon monoxide is "reforming." There are a number of methods to reform fuel. The three most commercially developed and popular methods are 1) steam reforming (SR), 2) partial-oxidation reforming, and 3) autothermal reforming (ATR). Each of these methods can be used to produce a fuel suitable for the fuel cell.

Steam reforming provides the highest concentration of hydrogen and can obtain a high fuel processing conversion efficiency. Partial oxidation (POX) is a fast process, good for starting, fast response, and a small reactor size. Noncatalytic POX operates at temperatures of approximately 1400°C, but adding a catalyst (catalytic POX or CPOX) can reduce this temperature to as low as 870°C. Combining steam reforming closely with CPOX is termed autothermal reforming.

TECHNOLOGY INTEGRATION WITH COAL GASIFICATION

As noted earlier, the numerous coal gasification systems available today can be classified as one of three basic types: (a) moving bed, (b) fluidized bed, and (c) entrained bed. All three of these types utilize steam, and either air or oxygen to partially oxidize coal into a gas product. The moving-bed gasifiers produce a low-temperature (425 to 650°C; 800 to 1200°F) gas containing devolatilization products such as methane and ethane, and a hydrocarbon liquid stream containing naphtha, tars, oils, and phenolics. Entrained-bed gasifiers produce a gas product at high temperature (>1260°C; >2300°F), which essentially eliminates the devolatilization products from the gas stream and the generation of liquid hydrocarbons. In fact, the entrained-bed gas product is composed almost entirely of hydrogen, carbon monoxide, and carbon dioxide. The fluidized-bed gasifier product gas falls between these two other reactor types in composition and temperature (925 to 1040°C; 1700 to 1900°F).

The heat required for gasification is essentially supplied by the partial oxidation of the coal. Overall, the gasification reactions are exothermic, so waste heat boilers often are used at the gasifier effluent. The temperature, and therefore composition, of the product gas is dependent upon the amount of oxidant and steam, as well as the design of the reactor that each gasification process utilizes.

Gasifiers typically produce contaminants that need to be removed before entering the fuel cell anode. These contaminants include H_2S, COS, NH_3, HCN, particulate, and tars, oils, and phenols. The contaminant levels are dependent upon both the fuel composition and the gasifier employed. There are two families of cleanup that can be utilized to remove the sulfur impurities: hot and cold gas cleanup systems. The cold gas cleanup technology is commercial, has been proven over many years, and provides the system designer with several choices. The hot gas cleanup technology is still developmental and would likely need to be joined with low-temperature cleanup systems to remove the non-sulfur impurities in a fuel cell system. For example, tars, oils, phenols, and ammonia could all be removed in a low-temperature water quench followed by gas reheat.

A typical cold gas cleanup process on an entrained-bed gasifier would include the following subprocesses: heat exchange (steam generation and regenerative heat exchange), particulate removal (cyclones and particulate scrubbers), COS hydrolysis reactor, ammonia scrubber, acid gas (H_2S) scrubbers (Sulfinol, SELEXOL), sulfur recovery (Claus and SCOT processes), and sulfur polishers (zinc oxide beds). All of these cleanup systems increase system complexity and cost, while decreasing efficiency and reliability. In addition, many of these systems have specific temperature requirements that necessitate the addition of several heat exchangers or direct contact coolers. For example, a COS hydrolysis reactor needs to operate at about 180°C (350°F), the ammonia and acid scrubbers need to be in the vicinity of 40°C (100°F), while the zinc oxide polishers need to be about 370°C (700°F). Thus, gasification systems with cold gas cleanup often become a maze of heat exchange and cleanup systems.

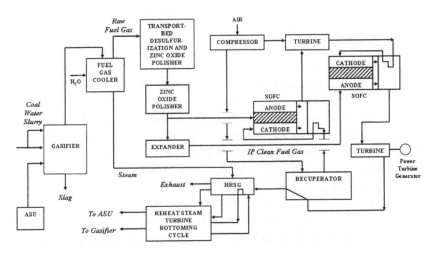

Figure 6.6 Schematic of 500 MW class coal fueled pressurized SOFC. Reconstituted from *Fuel Cell Handbook*, 5th ed., Oct. 2000, USDOE.

Figure 6.6 shows the coal-fueled solid oxide fuel cell power system based on work performed for the Department of Energy's Vision 21 Program to develop high-efficiency, low-emission, fuel-flexible processes (*Fuel Cell Handbook*, 2000). This study was based on a Destec gasifier, cascaded SOFCs at two pressure levels, an integrated reheat gas turbine, and a reheat steam turbine bottoming cycle configuration. The overall cycle net HHV efficiency is 59%, and is very near the 60% Vision 21 goal.

HYBRID SYSTEMS

Advanced power generation cycles that combine high-temperature fuel cells and gas turbines, reciprocating engines, or another fuel cell are the hybrid power plants of the future. As noted, these conceptual systems have the potential to achieve efficiencies greater than 70% and projected to be commercially ready by the year 2010 or sooner. The hybrid fuel cell/turbine (FC/T) power plant will combine a high-temperature, conventional molten carbonate fuel cell or a solid oxide fuel cell with a low-pressure-ratio gas turbine, air compressor, combustor, and in some cases, a metallic heat exchanger. The synergistic effects of the hybrid fuel cell/turbine technology will also provide the benefits of reduced greenhouse gas emissions. Nitrous (emissions will be an order of magnitude below those of non-fuel cell power plants, and carbon monoxide emissions will be less than 2 ppm. There will also be a substantial reduction in the amount of carbon dioxide produced compared to conventional power plants.

The hybrid system is key to the DOE's Vision 21 plants. The Vision 21 program has set power plant goals of achieving efficiencies greater than 75% (LHV) for natural gas. The higher efficiencies play a key role in reducing emissions, another target in Vision 21 plants. As a comparison, conventional coal-burning power plants are typically 35% efficient, and natural gas-fired plants are now 40 to 50% efficient.

The combination of the fuel cell and turbine operates by using the rejected thermal energy and residual fuel from a fuel cell to drive the gas turbine. The fuel cell exhaust gases are mixed and burned, raising the turbine inlet temperature while replacing the conventional combustor of the gas turbine. Use of a recuperator, a metallic gas-to-gas heat exchanger, transfers heat from the gas turbine exhaust to the fuel and air used in the fuel cell.

There can be many different cycle configurations for the hybrid fuel cell/turbine plant. In the topping mode described above, the fuel cell serves as the combustor for the gas turbine, while the gas turbine is the balance of plant for the fuel cell, with some generation. In the bottoming mode, the fuel cell uses the gas turbine exhaust as air supply, while the gas turbine is the balance of plant. In indirect systems, high-temperature heat exchangers are used.

Initial systems will be less than 20 MW, with typical system sizes of 1 to 10 MW. Future systems, in the megawatt class size, will boost efficiency even further by combining two solid oxide fuel cell modules with more advanced gas turbines and introducing sophisticated cooling and heating procedures.

Hybrid plants are projected to cost 25% below comparably sized fuel cells and be capable of producing electricity at costs of 10 to 20% below today's conventional plants. Operation of the plant is almost totally automatic. Therefore, it can be monitored and managed remotely with the possibility of controlling hundreds of power plants from a single location.

FUEL CELL TECHNOLOGY AND SYSTEM INTEGRATION ISSUES

There are no basic technological demonstration activities necessary for PAFC, only long life performance has to be confirmed. The performance and life expectancy of PEMFC is sufficient for domestic cogeneration. The major barrier for implementation are the high costs. In the longer term the development of cheap manufacturing methods and the reduction of the Pt catalyst load or the sensitivity for CO could be targets. Many more areas of technical improvement exist for high-temperature fuel cells. The expected high efficiencies cannot be achieved at the moment due to fundamental technical problems. Disadvantages of the MCFC are the very corrosive electrolyte, the low current density, and the current use of an external reformer. Material research is needed to find ways to reduce the corrosion and increase the current density. Another area of improvement is the development of modular reformers, which can easily be mass produced. Another option is to build a fuel cell network, where the fuel is processed in a central reformer and is fed in a pipeline network to a number of fuel cells near the user.

Problems with internal or external reforming are the same for SOFC and have to be solved in the same way. Additionally, a disadvantage for a simple SOFC system is the high operation temperature, which leads to problems for the auxiliary parts, such as heat exchanger, piping, and pumps. Another special SOFC problem is the use of brittle ceramics, which are difficult to handle. Some technologies like PAFC and PEMFC have reached commercial or pre-commercial stage at least in some applications; other technologies like the more advanced FCs (SOFC and MCFC) need at least 10 years before they are widely accepted as commercial technologies for large-scale stationary applications. Time and

R&D efforts are required in order to solve both technical and economic problems (for example, short life of some components, stress at high operating temperatures, high costs) and for better integration of all components in the plant.

Another critical parameter is the efficiency rate, which varies according to the type of cell. Electric efficiencies are usually significantly higher when pure hydrogen is used as fuel, which limits the feasibility of integrating coal gasification with fuels cells to those processes that can economically recover H_2. But also, overall system efficiencies for those fuel cells operating at high temperatures may be higher, due to the possibility of heat recovery produced by the electrochemical reaction. In general for PAFC, SOFC, and MCFC it is between 65 and 85%.

Areas for Technical Development

In general, technical developments will lead to a decrease in overall costs of this technology per unit of installed generating capacity (kWe). Some types of fuel cells need to achieve higher power densities per kg weight or m^3; most need to increase lifetimes of stacks or other plant components. For smaller applications, the technology must reach a reliable level sufficient to allow the plants to operate unattended.

The following is a list of specific development requirements for each fuel cell technology.

PAFC. The weaknesses of this technology include the need to limit as much as possible the presence of sulfur in the fuel feed and in the reformate stream, in order to avoid catalyst poisoning and frequent corrosion problems at the primary cell stack coolant circuit, and the need to increase electrical generation efficiencies of cell stack assembly lifetime to at least 40,000 hours (5 years); furthermore, an increase in power densities is an area where technical development is necessary (see Lane et al., 1995).

PEMFC or SPFC. For these technologies, areas needing further research are:

- Intolerance of the catalyst to CO and SO_2
- Ammonia production control

- Decreased cost, thickness, and weight of bipolar plates
- Membrane improvement (particularly in its ability to withstand dehydration) and improvement of membrane manufacturing techniques
- Improvement in hydrogen utilization from the reformate flow
- Investigation of alloying components and their effect on over voltage

MCFC. Among problem-areas, the following deserve special attention:

- Anode creep and sintering under compression
- Cathode dissolution (NiO) and reduction of dissolved nickel to a metal particle precipitate in the electrolyte matrix
- Corrosion of steel separator plate on the anode side
- Electrolyte migration in the external manifold
- Pressure optimization
- Selection of reforming type (external, direct internal, and indirect internal)
- Use of different fuels (coal mine methane, naphtha)

SOFC. Further research on this type of fuel cells is especially needed on:

- Manufacturing technology for the ceramic components (especially firing of the thin flat plates)
- Internal versus external reforming
- Operation data collection for a number of critical variables
- Control procedures for stack operation

LARGE-SCALE DISTRIBUTED POWER, INDUSTRIAL COGENERATION, AND CENTRAL GENERATION

It will be some time before fuel cells become available as a realistic commercial alternative for power generation applications. In order to compete with modern gas turbines/combined-cycle equipment, fuel cell plants will have to produce high-temperature waste heat, operate at an electrical efficiency of well over 50%, and cost below \$1000/kW. The fuel cells that may eventually be able to complete in these applications are the high-temperature fuel cell technologies that use molten carbonate or solid oxide (MCFC and SOFC) electrolytes. These complex technologies still have to overcome a number of technical hurdles and prove themselves at the system level.

One project under development at the NETL (http://www.NETL.DOE.gov) is an advanced power plant system that combines a multi-staged fuel cell with an extremely efficient turbine. Preliminary estimates show efficiencies greater than 80% (LHV). Studies showed that natural gas-to-electricity LHV efficiencies could break through an 80% barrier, while remaining cost competitive for a 4 MW solid oxide plant (tubular or planar). The Advanced Fuel Cell concept directly coincides with the long-term goals of the 21st Century Fuel Cell Program. These include system costs of \$400/kW and efficiencies of 70 to 80% or more (LHV to AC electricity), with fuel flexibility and a stack-life of 40,000 hours. They are intended for commercial application in 2015, maintaining ultra-low emissions.

Countries around the world are developing interest in the high-efficiency hybrid cycles. A 320 kW hybrid (SOFC and gas turbine) plant will entered service in Germany in 2001, operated by a consortium under the leadership of RWE Energie AG. This will be followed in 2002 by the first 1 MW plant, which will be operated by Energie Baden-Wurttemberg AG (EnBW), Electricite de France (EDF), Gaz de France, and Austria's TIWAG.

GASIFICATION TECHNOLOGY DEVELOPMENT AND SYSTEM INTEGRATION ISSUES

The key to successful commercialization of gasification systems devoted to power production rests both on maintaining the government R&D focused on industry needs and ensuring that the proper key technical issues are being addressed. The sidebar discussion summarizes those factors that have an impact on the future development of the technologies described here.

The short-term, prior 2010, markets and drivers that impact IGCC development and commercialization are:

- A function of steady, incremental improvements in system and unit economics and operation
- Affected by the price/outlook supply for NG, as well as portfolio diversification requirements (such as asset hedging against volatility issues among power producers)
- The materialization of niche market opportunities. These in turn will depend on:
  - Project economics, feedstock availability, regulation, and product integration
  - Low-cost feedstocks (e.g., petroleum coke/residue)
  - Marketable products (electricity, hydrogen, chemicals, other)
- Overseas markets, particularly outside of Western Europe, where infrastructure, project development schedules, and financing are uncertain

The successful transition to long-term markets (beyond 2010), which will ensure long-term mainstream deployment of technologies depends upon:

- Continual improvement of economy of scale, scheduling, and performance parameters, which will increase investor confidence
- Significant improvements in reliability and availability of technologies under development
- Demonstration of environmental and economic performance for disposition of hazardous wastes
- Streamlining of the environmental regulatory process
- National grassroots education programs
- Successful development next generation gasifiers.

The long-term markets are expected to show growth in two primary areas: clean power generation and clean energy conversion. Clean power generation refers to repowering; nearly 50% of the existing coal fleet is over 25 years old and require major overhauling.

Clean energy conversion refers to:

- Co-production of multiple products, e.g., fuels, chemicals, power
- Multiple feeds, e.g., pond feeds, sludge, biomass, industrial waste

In addition, there will be a transition to a hydrogen-based economy. This market timeframe is most likely to see central and distributed generation integrated with fuel cell technologies.

The highest priority technical issue that developing gasification technologies face is reliability. The key issues are:

- Present processes are unable to meet performance milestones on which economics are based.
- Single train availabilities must be at least 90% for utility applications and greater than 97% for refineries and petrochemical complexes.
- Power producers in general are unwilling or unable to guarantee integrated performance and risk large liquidated damages.
- There is the need to phase out multiple trains in order to improve economics.
- Both standardization and modularization of plants are needed in order to optimize costs and provide schedule and performance guarantees.

Plant reliability problems are associated with excessive downtimes due to component failure or scheduled maintenance requirement. Development needs include:

- Improved feed injectors to extend injector life from typically 2 to 6 months to more than 1 year.
- Improvements to refractories in order to reduce material cost and extend life from typically 6 to 18 months to more than 3 years. The need for refractories might be eliminated through new or novel gasifier concepts.
- More reliable and durable instrumentation. Temperature monitoring instrumentation needs to have life extensions beyond current 30 to 45 days. Furthermore, automated on-line feed (for fuel switching purposes) and on-line product analysis instrumentation are needed.

Key technical issues facing gasifier technologies on the whole evolve around economies of scale and the fact that most technologies are not suitable for low-rank coals. The primary developmental needs are:

- Small-scale gasifiers for distributed generation market, including hydrogen production
- Eliminating the need for costly and short-lived refractory linings
- Feedstock flexibility, focusing on adapting processes that are suitable for low-rank coals, high ash materials, biomass, and others
- Development of a new air separation plant, which greatly reduces cost and process complexity

Feedstocks pose challenges to mainstream integration of technologies. Three prominent issues are:

- The availability of alternative feedstocks; in particular, definitions are needed as to how much and at what costs.
- The cost and impact on injectors are key concerns for feedstock preparation. There are difficulties in injecting alternative materials into high-pressure gasifiers.
- Slurry feed systems raise serious concerns because of the thermodynamic penalty, thereby impacting on overall efficiencies as well as the cost of electricity and other products.

The primary developmental needs to address feedstock issues are:

- Development of reliable, continuous dry feed systems.
- Selection and demonstration of additives that will increase coal concentration in slurries.
- Development of reliable flow measurement and on-line analytical instrumentation.
- Developing new approaches to preparing and feeding low-rank coals and alternative feeds such as biomass. This includes addressing the removal of moisture from certain alternative fuels like biomass in order to increase energy content.

- Improved feedstock characterization is needed in order to accurately and reliably assess impacts on system performance.

Gas separation (i.e., air separation) is currently based on cryogenic technologies. Oxygen production represents up to 15% of the capital costs of IGCC plants and consumes 10% of the gross power production. Current technologies operate at either very high or very low temperatures. There is a great need for "step-out" technologies that have considerably lower capital costs and are more efficient. Ceramic-based membranes are currently viewed as favorable candidates. These have a high-temperature operability (typically > 850°C); however, they are still considered high risk. Intermediate operating temperatures for air separation units (i.e., 50 to 350°C) may be desirable, but currently such air separation technology is not available.

Hydrogen and CO_2 separation technology has challenges just as formidable as air separation. Three primary issues focus around:

- CO_2 separation, storage or sequestrations and/or utilization technologies must be demonstrated. Of highest priority is the need for storage technology before CO_2 removal can be even considered.
- Current technologies are restricted to either high-pressure H_2 or CO_2, but not both.
- Current technologies are restricted to operate at either very high or very low temperatures.

R&D needs must therefore focus on developing processes that operate within the preferred temperature range (<800°F), in the development of "step-out" technologies with lower capital costs and higher efficiencies, and in the development of processes capable of producing both high-pressure H_2 and CO_2.

The next R&D priority area concerns the syngas cleanup. It is important to recognize that industry has not invested in improving existing technologies or developing new concepts for gas cleanup in over 40 years. Existing technologies experience problems with or cannot remove certain trace contaminants in syngas. This creates problems for turbines as well as impacts on the reliability and life expectancies of FCs. So-called "deep-cleaning" technologies are both costly and inefficient, and are typically only used when high-value products are produced. R&D focus must be directed at addressing the following needs:

- Deep cleaning technologies are required in order to meet future environmental regulations for SO_X, NO_X, ammonia, hazardous air pollutants (HAPs), carbonyls, Hg, As, and others.
- The costs for such technologies must be equal to or lower than today's conventional technologies.
- New technologies that operate closer to downstream process requirements are needed. This means technologies that operate within the ranges of gas turbines and syngas conversion processes (i.e., 300 to 700°F).

Byproducts themselves have R&D issues. The two most critical ones are:

- Disposition of ash/slag and sulfur will become an issue as more plants are deployed using high-sulfur feedstocks.
- The impact of alternative feeds on ash/slag quality/ marketability needs to be assessed.

Specific needs to address these issues include low-cost technologies to enhance ash/slag quality (i.e., beneficiation); recovery of high-value trace metals from ash/slag; new market applications for ash/slag and sulfur; and new environmental test methods for each market application to ensure public safety.

Design standardization and modularization requirements mentioned early on concern four important issues:

- Very large plants will be needed for economic reasons. These are going to be difficult to finance.
- Each new project essentially must be designed from the beginning. This means costly design and engineering for each plant.

- Experience has shown plant start-ups to be difficult and that it takes time for capacities to reach critical mass.
- Low plant reliability plagues current technologies.

These issues need:

- Standardized/modularized plant designs, which can only come from well-established phased construction projects
- Value engineering tools, which must be developed in order to reduce costs and footprints
- Economic and efficient moderately sized plants

The last need translates into the requirement that plant sizes need to be consistent with the needs of utilities and other industries. To date, a detailed market study that clearly defines this need does not exist. Also, economy and efficiency means that plants that can be mass-produced and deployed on a modular basis must be developed. Finally, start-up and reliability problems can only be overcome by standardizing operations, which, as noted early on, will increase both investor and customer confidence.

Three additional R&D support areas should be noted. These are areas that address issues concerning synthesis gas utilization, issues surrounding informational databases, and requirements for advanced instrumentation. In terms of synthesis gas utilization, we must recognize that gasification technology is driven in part by the natural gas market. There are real concerns about future availability and development of gas turbines and/or natural gas due to various market forces. Coupled with this are uncertainties concerning the long-term impact of contaminants in the syngas. To address these issues, R&D needs to focus on both gas turbines and fuel cells. For gas turbines, optimized development of syngas combustion and a better understanding of the effect of impurities on gas turbine performance are among the top priorities. Impurities in the gas can impact turbine performance by causing delamination, spalling, embrittlement, and the deterioration of thermal barrier coatings. Gas turbines also need low NO_x or catalytic combustors for syngas uses.

For fuel cells, the two top needs are a drastic reduction in cost and demonstrated tolerance for contaminants in the syngas.

Issues surrounding informational databases needed to bring technologies forward are the avoidance of repetitive mistakes among developers (information is not freely shared) and the missing of key performance milestones that result in liquidated damages to developers. Informational databases truly need an industry-wide information management system. The key components of such a system are reliable statistics on performance parameters for existing plants, and detailed component cost, availability, and reliability indices. Such an information system should be extended to feedstock performance, i.e., provide chemical and physical properties of various feedstocks with standard deviations and variances, and include a data/knowledge base on reactivity and gasification characteristics. A final component of such a system is a knowledge base and reliable indices that characterize operation and maintenance problems. This component should address generic O&M problems common to the technologies and industry itself. These in fact could be design practices such as a "slurry handling design manual."

For instrumentation, the issue is rapid, affordable, on-line instrumentation. The development of such tools is tantamount to further advancement of coal gasification. Specific needs include affordable on-line analytical devices to provide elemental composition of feedstock along with time varying composition and other properties, reliable on-line flow measurement, rapid on-line analysis of slag viscosity along with instrumentation that measures slag thickness on refractories, on-line instrumentation to track refractory wear and damage, on-line product gas analysis (which should include detection of trace components), and isokinetic particulate measurement capabilities, preferably in the gasifier itself.

RECOMMENDED RESOURCES

1. A.G. Di Mario, F. (1994) Prospettive di Sviluppo e Applicazione dei Sistemi con Celle a Combustibile. *Ambiente, Risorse, Salute.* July/August.

2. Arthur D. Little (1991): Prospects for Commercialization of Fuel Cells: A report to The Platinum Association and to Mannesmann Plant Construction.

3. Barp, B. Designing a SOFC fuel cell system for household energy supplies, Proceedings, Fuel Cells in the Energy Market, Cologne, 12–14. March 1996.

4. Barthels, H., Brocke, W.A., Bonhoff, K., Groehn, H.G., Heuts, G., Lennartz, M., Mai, H., Mergel, J., Schmid, L., Ritzenhoff, P., PHOEBUS Jülich — An autonomous energy supply system comprising photovoltaics, electrolytic hydrogen, fuel cell, Hydrogen ENERGY Progress XI, Proceedings of the 11th World Hydrogen Energy Conference, Stuttgart, 23–28 June 1996.

5. Beckervordersandforth, Ch., An economic comparison of cogeneration systems, Proceedings, Fuel Cells in the Energy Market, Cologne, 12–14. March 1996.

6. BP Statistical Review of World Energy, June 1996.

7. Energieforschung und Energietechnologien, Annual report 94, funding, German Ministry of Education, Science, Research and Development (BMBF).

8. Escombe, F.M. (1995) Fuel Cells: Applications and Opportunities — Executive Summary. ESCOVALE Consultancy Services, Report n. 5020. January.

9. FUEL CELLS 2000 http://www.fuelcells.org

10. Fuel Cells and their Applications — Karl Kordesch, Guenter Simader (abstract, table of contents) — Copyright by VCH Verlagsgesellschaft mbH, D-69451 Weinheim, Federal Republic of Germany, 1996.

11. Geißler, E., Fuel Cells, Basis, History, Status, Internal Paper, Forschungszentrum Jülich, Nov. 1990 (in German).

12. Hagey, G., Marinetti, D., Mueller, E.A., Status of fuel cell system development/Commercialisation and future systems energy, environmental, and economic benefits, pre-prints: International Conference, Next Generation Technologies for Efficient Energy End Uses and Fuel Switching, Dortmund, 7–9. April 1992.

13. Hörmandinger, G., Fuel Cells in Transportation, Imperial College of Science, Technology and Medicine, London, Sept. 1995.

14. HYDROGEN & FUEL CELL LETTER http://www.spice.mhv.net/~hfcletter/

15. IEA Implementing Agreement on Advanced Fuel Cells, Annex V: Fuel Cell System Analysis, Final Report, May 1995.

16. IKARUS-Retrievalsystem, Instrumente für Klimagasreduktionen, Fachinformationszentrum Karlsruhe, Version 1.A.2.

17. Kartha, S.; Grimes, P., Fuel Cells: Energy conversion for the next century, Physics Today, Nov. 1994, pp 54–61.

18. Kasahara, K., Fuel Cells in Japan — A Survey, Proceedings, Fuel Cells in the Energy Market, Cologne, 12–14. March 1996.

19. Lane, R.M., M.R. Fry and J.N. Baker (1995) Initial Assessment of Fuel Cells. A Greenhouse Gas R&D Programme. EA Technology. Report n. PH2/1.

20. Ledjeff, K. (Hrsg.), Anahara, R., PAFC Plants in Japan, Brennstoffzellen, Entwicklung, Technologie, Anwendung, C.F. Müller Verlag, Heidelberg, 1995.

21. Moser, Th., PEM fuel cells — development status and applications, Proceedings, Fuel Cells in the Energy Market, Cologne, 12–14. March 1996.

22. Nymoen, H., Knappstein, H., Ledjeff, K. (Hrsg), Kraft-Wärme-Kopplung mit Brennstoffzellen — Erste Betriebserfahrung mit 200-kW-PAFC, Brennstoffzellen, Entwicklung, Technologie, Anwendung, C.F. Müller Verlag, Heidelberg, 1995.

23. Nymoen, H., PAFC demonstration plants in Europe: first results, Proceedings of the third Grove fuel cell symposium, London, 28.9. — 1.10. 1993, Journal of Power sources, Vol. 49, pp 63–77.

24. Roesler, R., Zittel, W., 'Wasserstoff als Energieträger,' IKARUS-Instruments for reduction strategies for climatic gases, Sectoral project 4: 'Umwandlungssektor,' project management: Group Technology Assessment (TFF), Jülich 1994.

25. Second European Solid Oxide Fuel Cell Forum, Oslo, 6–10 May 1996.

26. Sjunnesson, L., Current status and development potential of fuel cells, Proceedings, Fuel Cells in the Energy Market, Cologne, 12–14. March 1996.

REFERENCES

1. Patel (1995). "Fuel Cell" Chapter XIII of the volume Overview of Energy R&D Options for a Sustainable Future. Contract JOU2-CT93-0280. European Commission DGXII.

2. http://fce.com

3. Ghezell-Ayagh et al. (2001) An explicit model for direct reforming carbonate Fuel Cell stack, *IEEE Trans. Energy Conversion*, Vol. 16, No. 3.

4. Ghezell-Ayagh et. (2001) Operation and control of direct reforming fuel cell power plant, *Proc. IEEE Power Eng. Soc. Winter Meeting.*

5. Ghezell-Ayagh et al. (1999) Development of a stack simulatioin model for control study on direct reforming molten carbonate fuel cell power plant, *IEEE Trans. Energy Conversion*, Vol. 14, No. 4.

6. *Fuel Cell Handbook*, USDOE, October 2000. 5th edition.

Index

T - #0095 - 101024 - C0 - 234/156/19 [21] - CB - 9780824722470 - Gloss Lamination